481 ILLUSTRATIONS IN COLOR

FOSSILS

A GUIDE TO PREHISTORIC LIFE

by

FRANK H. T. RHODES

President, Cornell University

HERBERT S. ZIM

PAUL R. SHAFFER

ILLUSTRATED BY

RAYMOND PERLMAN

Professor of Art, University of Illinois

GOLDEN PRESS • NEW YORK

Western Publishing Company, Inc.

Racine, Wisconsin

FOREWORD

This guide to life of the past differs from other Golden Nature Guides. Instead of dealing with a single group of plants or animals, it deals with them all. Instead of being concerned only with the immediate present, its scope covers over half a billion years. Instead of dealing with life first hand, this guide must rely on only scant clues—bits of shell, bone, or sundry fossil impressions. Such clues are scarce, so each must be studied minutely. Details are important and they have been stressed in the systematic survey of fossil forms. Most fossils have only scientific names and these often refer to groups rather than to species.

We have many institutions and individuals to thank for aid with this book. Dioramas and murals of the Chicago Nat. Hist. Museum are the basis for many of our restorations. The University of Illinois, the Illinois Geological Survey, the U.S. National Museum, the University College of Swansea, the Ward's Natural History Establishment, Inc. loaned us specimens. So did M. W. Sanderson, A. F. Hagner, W. W. Hay and F. J. Koenig. Angela Heath and Shirley Osborne have assisted the senior author, and many more of our colleagues have given us photographs, specimens, comments, and suggestions.

F.H.T.R.
H.S.Z.
P.R.S.

 Produced in the U.S.A. by Western Publishing Company, Inc. Published by Golden Press, New York, N.Y. Library of Congress Catalog Card Number: 62-21640

CONTENTS

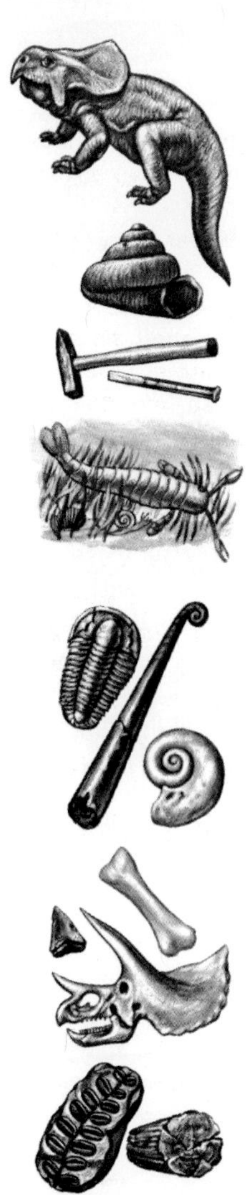

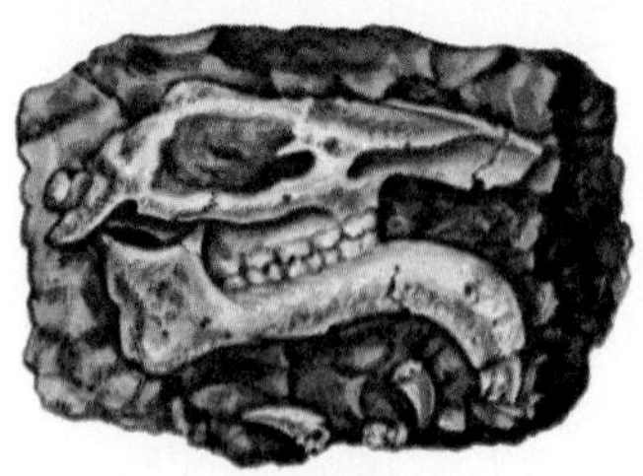

Dinotherium fossil

restoration

LIFE, PAST AND PRESENT

The earth teems with life. Mountains, prairies, deserts, beaches, lakes, rivers and seas—every part of land, sea and air is inhabited by living things. The number of different species of living things is enormous. More than 350,000 species of plants and 1,120,000 species of animals are known.

How did these many species originate? Has life always been the same as it is now? Men have asked these questions for thousands of years. To answer them we must turn to fossils and to a knowledge of living organisms and their structure. Only an understanding of living animals can put life in the fragments of bones and shells millions of years old.

The elephant is the largest living land animal. But the study of fossils shows not only that elephants are a recent group in the long history of living things but also that early elephants looked more like hogs. As geologists trace elephant fossils from older to younger rocks they piece together the history of elephant evolution. Fossil bones and teeth reveal the structure of early elephants, but by studying these fossils in the light of the anatomy of living elephants, complete reconstructions of extinct elephants can be made with reasonable accuracy. Some unusual occurrences of mastodon fossils with crude flint weapons prove that these elephants were hunted by our ancestors.

EVOLUTION OF ELEPHANTS

Living elephants are survivors of an ancient, more widespread and varied group, which evolved from pig-sized ancestors of the Upper Eocene. (See the geologic clock, pp. 30-31.) Only a few of the many extinct elephants and their kin are shown.

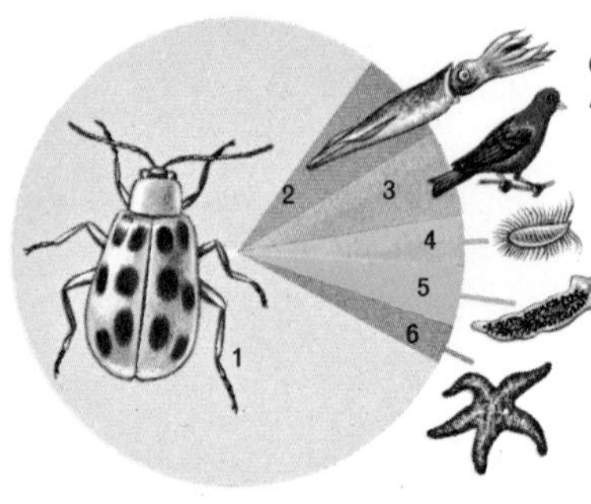

OVER 1,000,000 ANIMAL SPECIES

1. Arthropods—900,000
2. Mollusks—45,000
3. Chordates—45,000
4. Protozoans—30,000
5. Worm-like phyla—38,000
6. Other invertebrates—21,000
   Approximately 1,000,000

ALL FORMS OF LIFE have evolved from early beginnings, some three billion years ago. From relatively few primitive forms, the major groups of plants and animals developed. Living things became more complicated and adapted to many different ways of living. The number of different species gradually increased until they reached the tremendous diversity of today. The study of fossils (paleontology) traces the various paths by which animals and plants evolved to their present forms. Some, like elephants and horses, have changed greatly through the ages. Others, like the horseshoe crab and cockroach, have not changed in hundreds of millions of years. Still other fossils show lines of development that came to a dead end. Giant Sloths, once plentiful, are known only as fossils.

GLYPTODONT, 9 ft., an armored mammal from the late Cenozoic, is a fossil that shows spectacular and obvious adaptation. This relative of the armadillos was protected against carnivores and other enemies by a thick, solid, domed armor, which reached 5 ft. in length in some forms. The head and tail were also armored, and in some species the tail terminated in a spiked, mace-like club. Yet despite, or because of, these unusual adaptions glyptodonts became extinct.

ABOUT 350,000 PLANT SPECIES

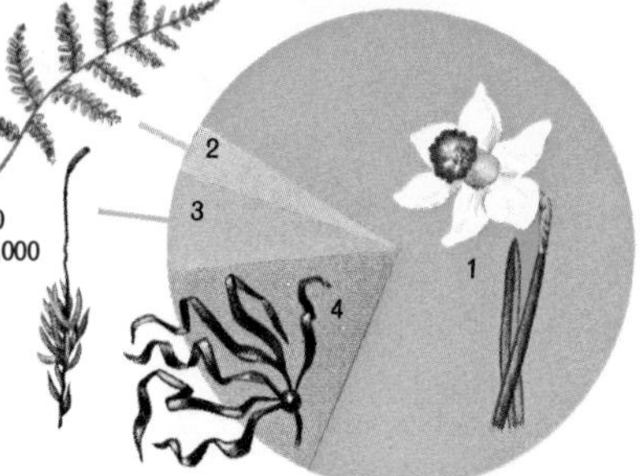

1. Flowering Plants—250,000
2. Ferns, Conifers, etc.—10,000
3. Mosses and Liverworts—23,000
4. Algae, Fungi, etc.—60,000

About 350,000 plant species

ADAPTATION Most plants and animals exist only because they are successfully adapted to their environments. Each distinct environment such as a desert, pond or mountain top supports a more or less distinct population of animals and plants. Those which, over long periods of time, have become fitted to cope with local conditions have survived. All the rest have become extinct. Many living things are uniquely adapted to particular environments. The streamlined shape of a fish and the structure and function of its fins and tail are adaptations to life in the water. The fleshy stems of a cactus are adaptations that conserve water in the desert. Such adaptations succeeded, but the fossil record is strewn with the remains of those that failed. The slogan of life may well be—adapt or perish.

Survival in animals depends on adaptations as varied and as intricate as the animals themselves. Virtually every structure of a plant or animal may be regarded as adaptive. Many animals have protective coloring and a few forms, such as the bottom-living flounder, are able to change their color to conform to their background. Such an intricate adaptation is rarely discernible in fossils. However, if the adaptation affects bone or shell, it may show up clearly in the fossil record.

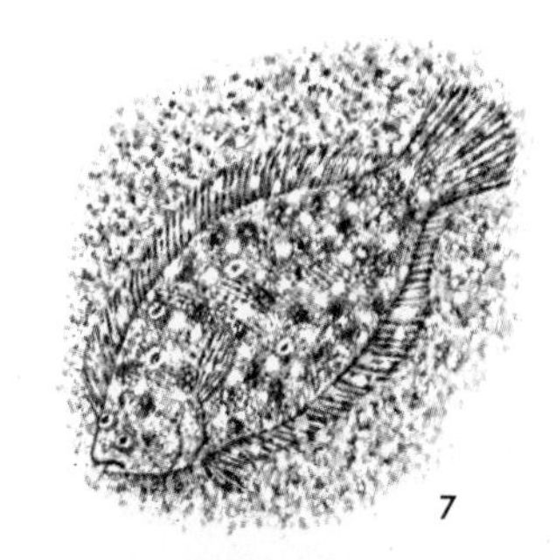

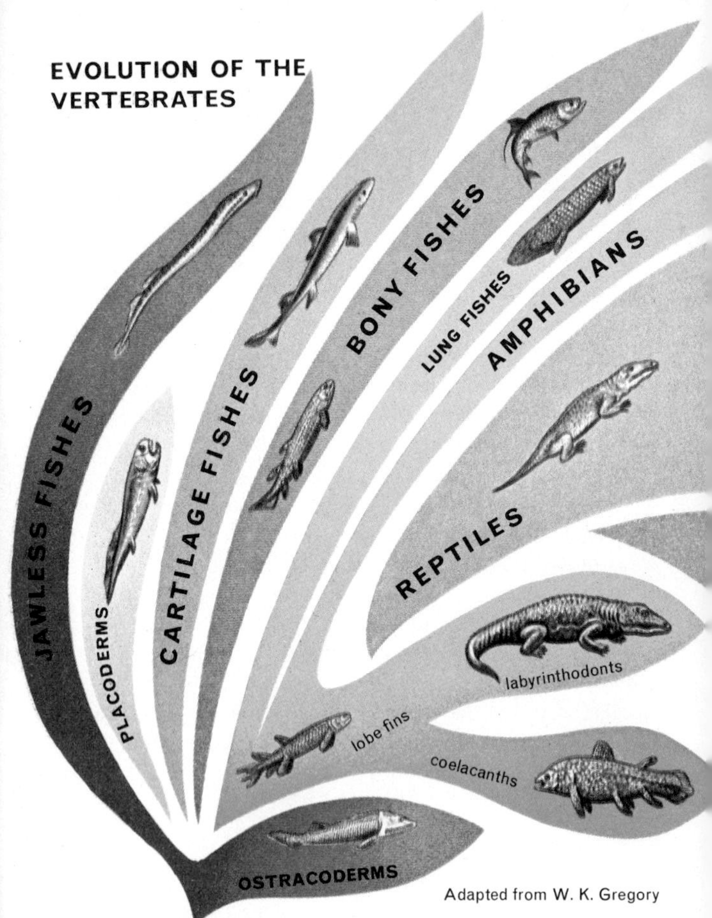

Adapted from W. K. Gregory

DEVELOPMENT OF MODERN ANIMAL LIFE is difficult to trace because the fossil record is incomplete. Enough is known to suggest the general pattern of evolution and to reconstruct in some detail history of groups in

those areas where fossils occur abundantly. The chart shows relationships among major groups of vertebrates. Those animals within a given color probably developed from common ancestors.

A Woolly Mammoth partly preserved in Siberian frozen ground. Discovered in 1900. Ht. 10 ft.

INTRODUCTION TO FOSSILS

Fossils are the remains of prehistoric life or some other direct evidence that such life existed. To become fossilized a plant or animal must usually have hard parts, such as bone, shell, or wood. It must be buried quickly to prevent decay and must be undisturbed throughout the long process. Because of all this very few plants or animals that die are preserved as fossils.

In rare cases whole animals may be preserved. In Siberia and Alaska fossil mammoths have been found in the frozen ground, completely refrigerated for some 25,000 years. In Galicia (Poland) an Ice Age Woolly Rhinoceros was well preserved in asphalt. In semi-arid South America, parts of mummified Ground Sloths have been preserved in caves. In each of these cases an unusual condition—cold, chemical action and dryness—was involved.

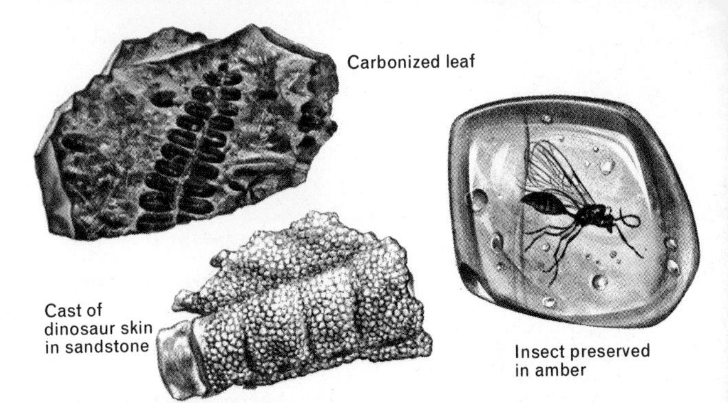

SOFT PARTS are rarely found intact but insects' exoskeletons and minute appendages have been preserved in amber, the hardened resin of ancient trees. Leaves and small, soft marine animals buried in mud which hardened into shale have sometimes left behind a thin film of carbon outlining their form and preserving delicate details of their structure. And, in western Canada, sandstone casts of dinosaur skins have been preserved.

HARD PARTS are often preserved with little or no alteration. Teeth of sharks and mammals are examples, and small jaws of ancient sea worms have been found. Bones may be preserved but more often have been altered and replaced by dissolved mineral matter. Shells frequently remain intact and in a few places logs and stumps have been preserved in peat or coal.

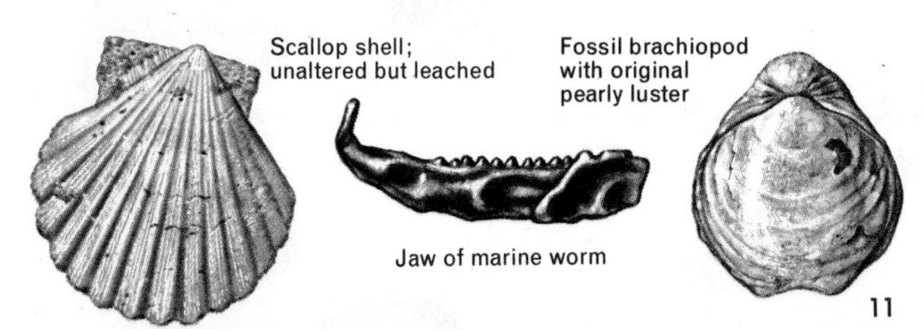

Petrified Forest, Arizona

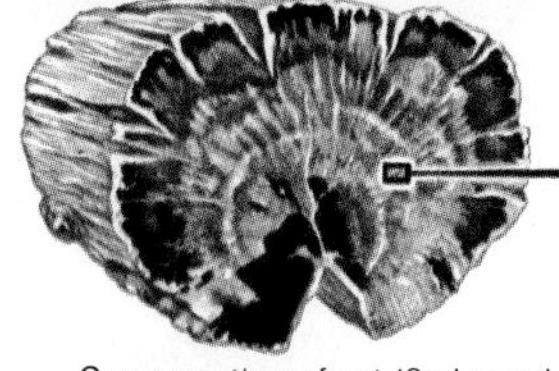

Cross section of petrified wood

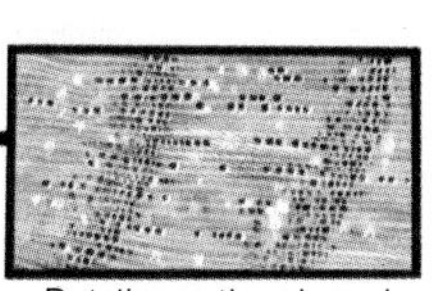

Detail, greatly enlarged, showing cells

ALTERATION of hard parts preserved as fossils is common. Circulating water dissolves chemicals from shells and bones and leaves them light and spongy. More often as chemicals are dissolved they are replaced by others. Silica, lime and iron compounds are commonly deposited in fossils. Sometimes this replacement preserves the original structure of the plant or animal completely. In some petrified wood, silica has replaced the original woody structure so perfectly that the cells and annual rings show clearly. In most petrified wood and most replacement fossils, the replacement is less perfect and shows only the general form.

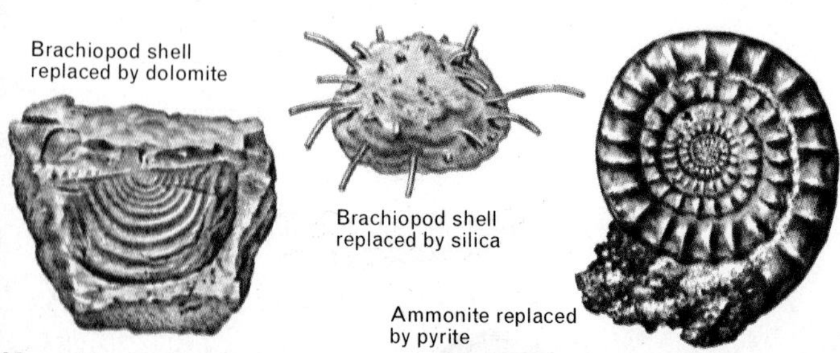

Brachiopod shell replaced by dolomite

Brachiopod shell replaced by silica

Ammonite replaced by pyrite

1. Animal dies and shell is buried in the sand.

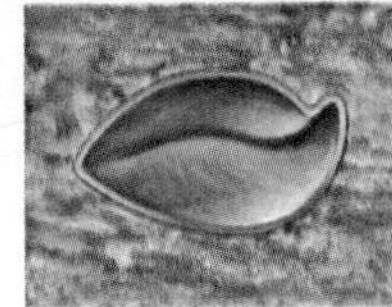

2. Sand hardens to rock. Shell interior unfilled.

3. Shell material dissolved. Cavity wall is mold of shell.

4. Dissolved chemicals fill mold to form cast.

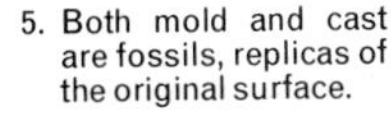

5. Both mold and cast are fossils, replicas of the original surface.

MOLDS AND CASTS Not all fossils are bones, shells and other remains. Some are mere indications of prehistoric life. All the original plant or animal material may be dissolved away so that only a cavity remains—the walls of which are a natural mold of the fossil. Later, dissolved substances may fill the cavity, forming a natural cast of the original. Such casts are common fossil forms. Footprints or trails of animals may harden as a mold. Filled with fresh mud, casts are formed and both may be preserved, as in the red sandstone (Triassic) of the Connecticut Valley which contains tracks of dinosaurs.

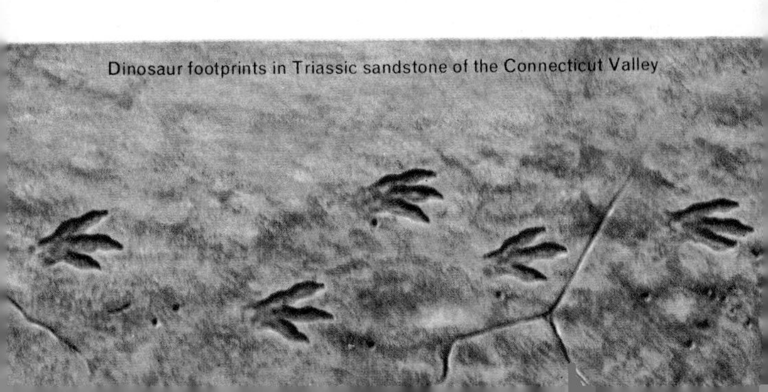

Dinosaur footprints in Triassic sandstone of the Connecticut Valley

mollusk borings

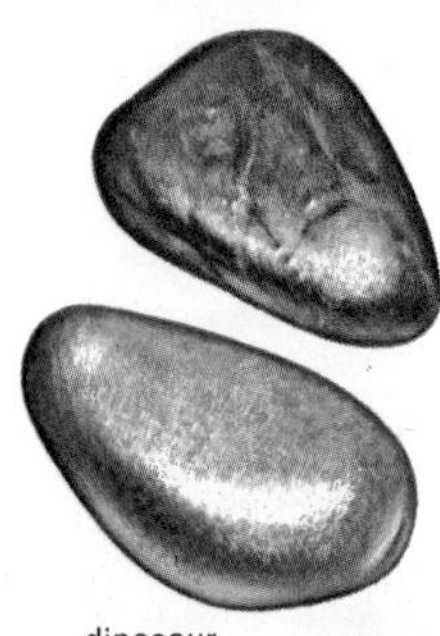

dinosaur gastroliths

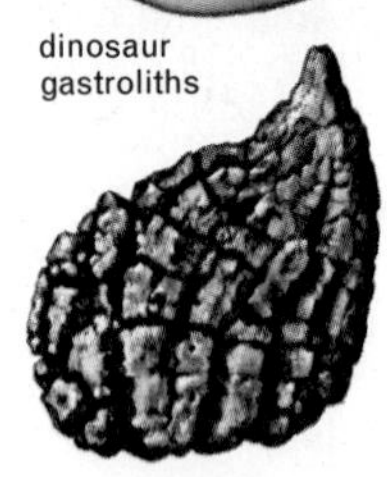

coprolite

Aurignacian hand axe

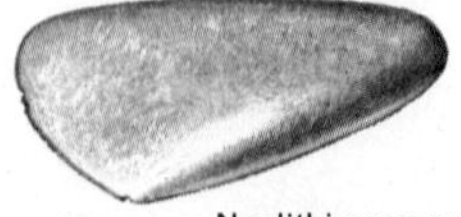

Neolithic scraper

OTHER TYPES OF FOSSILS include some curious forms, all of which are evidence of ancient life.

BORINGS of worms and mollusks indicate that these animals lived millions of years ago. Such fossils are common. Sometimes petrified wood shows borings also.

GASTROLITHS are smooth, rounded pebbles found in rib cages of dinosaurs. These stones probably aided the dinosaurs' digestion just as gravel in their gizzards helps chickens crush grain. Polished gastroliths are found only in "dinosaur country."

COPROLITES are fossil excreta and give a clue to the diet of ancient animals. These lumpy fossils are usually associated with land animals of the past 50 million years.

ARTIFACTS are stone tools or weapons made by ancient man. Found in many parts of the world, the oldest have been found with bones of animals now extinct. The first stone artifacts were crude and difficult to recognize. More recent ones were chipped and polished to make beautiful implements.

SEDIMENTARY ROCKS contain nearly all the fossils that are found. These rocks are formed of sediments—mud, sand, clay—deposited mechanically, chemically or by organisms, in seas, lakes, caves, deserts and river valleys.

Grand Canyon strata

STRATA or layers are a characteristic of sedimentary rocks. The bottom layers are naturally the oldest. But not all sediments are evenly or clearly bedded.

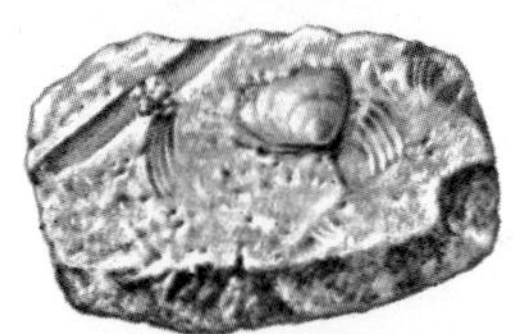

limestone

LIMESTONE, mainly calcium carbonate, common in warm, shallow seas, often has fossils.

shale

SHALE is a fine-grained rock formed from silt and clays. It preserves fossils well.

SANDSTONE is widespread in desert deposits and in shallow water sediments.

sandstone

RIPPLE MARKS AND MUDCRACKS characterize many sedimentary rocks formed in shallow waters. Ripple marks are common in shale. Mudcracks may form as mud and clays dry. These imply the presence of sunlight, water and moderate temperatures—conditions related to the possibilities of life.

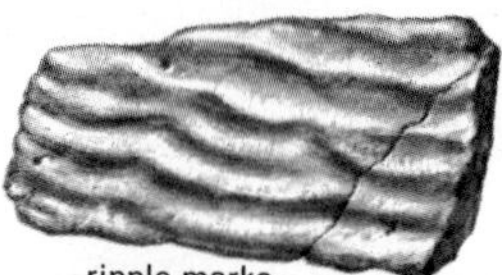

ripple marks

mudcracks

North American sedimentary formations deposited during the past 600 million years. Small local beds of sedimentary rock are also found within areas of igneous and metamorphic rock.

SEDIMENTARY ROCKS, often rich in fossils, occur over much of North America. But in many places the solid rocks are covered with soil or glacial deposits, or the fossil-bearing layers lie deep beneath other rocks. Hence fossil hunting is restricted to outcrops—places where the sedimentary rock is exposed at the surface, as in cliffs, river banks, roadcuts or quarries.

The fact that fossils are found in sedimentary rocks is no coincidence. Other rocks are subjected to forces or conditions which destroy fossils easily. The processes

Fossil bony fish from Eocene Green River formation, Wyoming.

that wear down the earth's surface produce the sediments from which sedimentary rocks are formed. These wearing down processes (degradation) involve rainfall, evaporation, wind, running water and transportation among other things. And the fossil fish pictured above not only proves that fishes lived in the distant past, but that conditions in the lake in which it lived were not greatly different from those in many areas today. This and older fossils provide evidence that basic physical conditions making life possible today existed not only 50 million years ago when this fish became a fossil but probably go back about two and a half billion years. Every fossil, even the most common, tells a fascinating story of the changing surface of the earth and the development of life upon it.

concretion

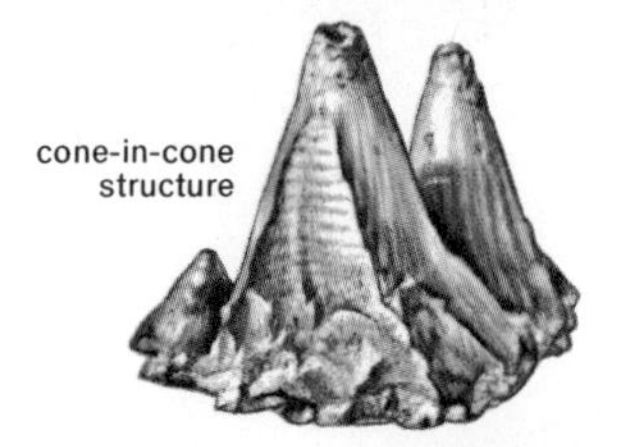
cone-in-cone structure

septarian nodule

dolomite "pseudo-coral"

Pseudofossils, often shaped like fossils, lack detailed fossil structures.

PSEUDOFOSSILS are rock structures that resemble fossils. They may have any shape and often look like parts of plants or animals. A geologist will usually recognize a pseudofossil at once, but an amateur may be misled. Pseudofossils resemble fossils only in external form. They never have the detailed structure of true fossils. They may occur in improbable situations, as for instance a "footprint" in rock formed long before any creatures walked on land.

Pseudofossils are formed in many ways. Some are water-worn fragments of rock. Concretions which form in sedimentary rock may contain a fossil, though most do not. Concretions, harder than the rock in which they occur, are often found on the surface. Some minerals form dendrites or fernlike deposits on or in rocks. Moss agates are dendrites, not fossil moss.

pyrolusite dendrites on dolomite

polished moss agate

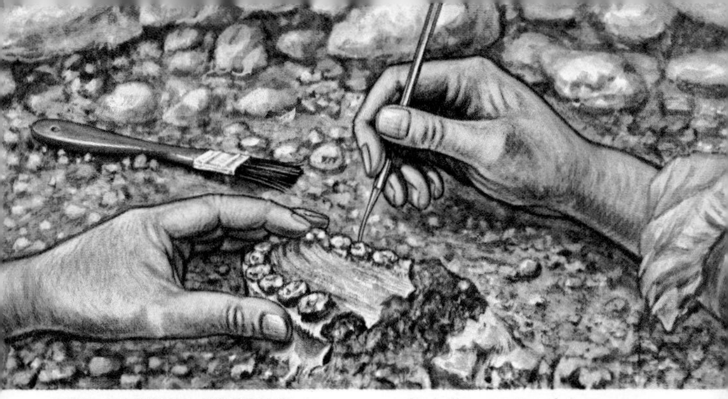

THE RAREST FOSSILS are those of human beings. This jawbone unearthed in Africa in 1961 pushed the origin of humans or near-human, tool-using animals back to nearly 1,750,000 years ago.

FOSSILS FOR AMATEURS

Collecting and studying fossils can be an interesting hobby as well as an important science. Only during the past two centuries has paleontology, the study of fossils, moved to the professional level. Amateurs have collected and studied fossils much longer and today they enjoy field trips and collecting as much as ever. Major discoveries have been made by amateurs and many have won acclaim from professional geologists.

Unless the ground is covered with snow, collecting fossils is an all-year occupation. It takes you out-of-doors and off the beaten track. You learn to know your region intimately and enjoy the company of other local "rockhounds." No other hobby can open such wide vistas of time and space. The study of fossils still has many unsolved problems which a serious amateur can tackle with some chance of personal success. Such a person will understand fossils better if he also keeps up a continued interest in living animals and plants.

FOSSILS RECONSTRUCT LIFE They enable scientists to picture accurately many kinds of long-extinct plants and animals.

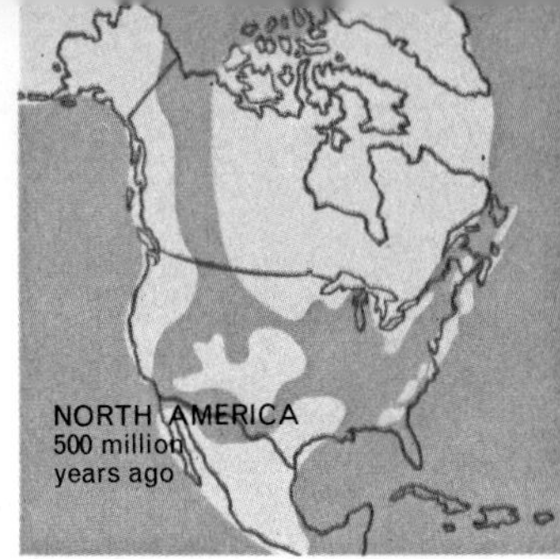

FOSSILS PLOT GEOGRAPHY They indicate ancient land and water areas and show the changing continents.

WHY COLLECT FOSSILS? Most people collect for the simple fun of it—for the fun of tramping and exploring; for the excitement of a rare find; for the challenge of "working out" a perfect specimen. But in the course of doing all this, the layers of sedimentary rocks unfold like pages of a gigantic book, revealing the fascinating story of the earth's long and exciting past. Events 50, 100 or 500 million years ago become real because the fossils you have found provide a clear connection with bygone ages.

With the aid of fossils the reconstruction of prehistoric plants and animals was possible, and the story of the evolution of life became clear. Without the evidence of fossils, evolution would still be a theory, not a fact. Fossils help determine whether sediments were formed in shallow or deep seas, in rivers, in swamps or in deserts. Thus they give a clue to the geography and ecology of the past and show how the continents and seas have changed. Fossils prove that Alaska was once connected with Siberia and Australia with Malaya. The distribution of shallow-water mollusks aids in tracing ancient shorelines.

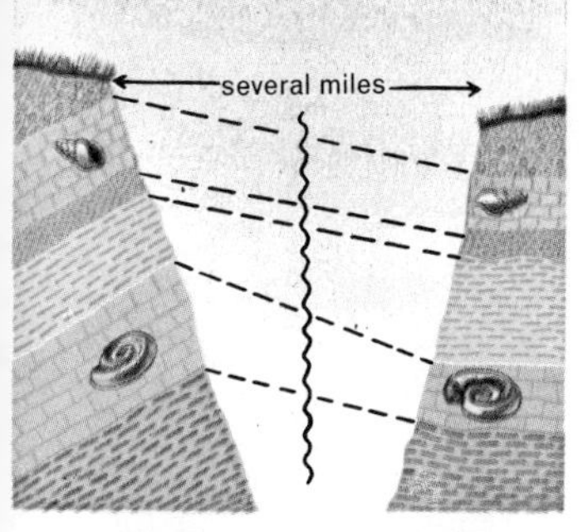

FOSSILS HELP CORRELATE STRATA Index fossils establish the time relationship between rocks of different areas.

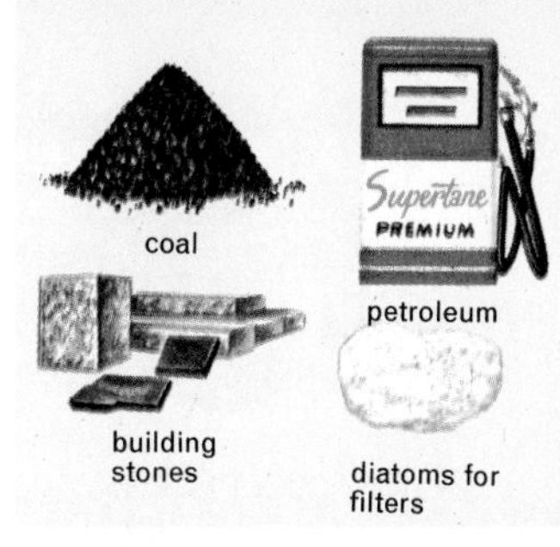

FOSSILS ARE NATURAL RESOURCES Fossils are a source of coal, oil, lime, phosphate and building stones.

Fossils, in addition to being clues to ancient geography, are also clues to the climate of the past. Fossil corals show that warm, shallow seas once covered New York. And plant fossils show that the climates of Antarctica and Greenland were once mild.

Certain fossils of limited time distribution clearly mark certain beds or strata of rocks. These are index fossils and their occurrence in rocks located miles apart proves these rocks were formed at the same time. This use of fossils to correlate strata is important in mapping rock formations and in locating valuable mineral deposits.

Fossils themselves or rocks located by fossils provide natural resources valued at billions of dollars. Nearly all our fuels are fossil fuels. Coal and oil are the remains of ancient plants and animals. Fossil limestones make excellent building stones. Some are cut for ornamental use. Micro-fossils are used as filters, fillers, in polishes and for many other purposes. Some phosphate beds are associated with large deposits of fossil bones. Amber and jet are fossils used as jewelry.

Rockhounds in a quarry

STUDYING FOSSILS is something like making a rabbit stew—you first must get a rabbit. Considering the earth as a whole, fossils are rare. Many are buried beneath the sea. Forests, grasslands, swamps, deserts, soil and rock debris cover many more. Yet despite all this, fossils are often easy to find.

WHERE TO LOOK for fossils is easily settled. Look in sedimentary rocks, for these are the principal rocks which may contain fossils. Occasionally fossils are found in beds of volcanic ash or are even preserved in lava, but these are rare. Sedimentary rocks (mainly sandstone, shale and limestone) are common, but not all of them contain fossils. Maps in this book show where such sediments are exposed but this rough data must be supplemented by detailed maps and state geological publications (see p. 25).

In general, fresh exposures of rock are best for collecting. Look in road or railroad cuts. Visit mine dumps, quarries and places where rock is being excavated for new construction. Cliffs, river banks, headlands and other natural exposures are good places also. Remember that all these places involve a certain element of danger. Watch for traffic at roadcuts and get permission before entering quarries. Loose rocks can be a danger to you and to anyone below you.

TOOLS for collecting fossils are similar to those used by any rockhound. A geologist's, plasterer's or brick-layer's hammer is essential. So is a knapsack or stout shoulder bag for carrying specimens. Fossils are often delicate. Take newspaper and wrap each specimen separately as soon as it is collected. Put a label or slip of paper with each specimen giving location, formation, date, and identification if known.

A large and a small cold chisel are needed to remove specimens, for hammering alone is rarely enough. A small shovel and a steel wrecking bar may also prove handy. You will often need road maps and more detailed topographic or geological maps to locate your outcrops. A magnifying glass or hand lens (5 to 10 power) is worth having. Carry a compass, a first-aid kit and a pocket knife. You may need to carry your own food and water too.

HOW TO COLLECT Take time to survey the area. Look for rock surfaces where weathering has exposed fossils. Weathered-out specimens are easy to collect and may be in excellent condition. Turn over rock fragments and study all sides. Break open concretions if they occur. Should you locate vertebrate bones or any fossil you believe rare, leave it intact and get professional help. Valuable fossils have been ruined by bungled attempts to remove them.

PREPARATION AND CLEANING should be done at home after specimens have been removed. The delicate task of cleaning and working out a specimen is best done on a stout table with good light and adequate tools. Old dental tools are excellent for this purpose. Electric-powered "hobby sets" often contain small drills and grinders which make the task easy. Bone and other delicate fossils may require a coating of a preservative such as shellac or Alvar to prevent cracking and deterioration.

IDENTIFICATION AND EXHIBITION of your collection bring your work to a climax. Regional volumes on fossils should be consulted for identification. Secure the aid of geologists at universities and museums. These experts are usually glad to aid an amateur. Spot your specimen with a drop of quick-drying enamel and put your catalog number in India ink on the spot. Record this number and the name, location and other data on a card and also in a separate catalog.

Fossils can be stored in cardboard trays of varying sizes purchased from scientific supply houses. Put your label in the tray beneath the specimen. Keep small specimens in vials. Build exhibit cases or a bank of shallow drawers to hold trays and specimens.

MAPS ARE IMPORTANT because so much information is presented in them about sedimentary formations, structures and strata. Without maps the location of fossil deposits is impossible; use all kinds.

Road maps: Obtain sets of several kinds covering your area. Keep them up to date. Mark small back roads that may not appear on the maps. Keep one set at home for reference. Take another set into the field with you.
Topographic maps are produced by the U.S. Geological Survey. These are more detailed than road maps and show the land and water features as well as culture (roads, bridges, towns, houses, etc.). An Index Map for your state can be obtained from the U.S. Geological Survey, Washington 25, D.C.
Geological maps: These are often prepared by state geological surveys either as a state geological map or in relationship to specific reports. Check with your state survey and get their list of publications.

BOOKS will help you identify fossils, understand basic geology and the story of evolution. Check the publications and reports of your state geological survey. Write the Supt. of Doc., Wash. 25, D.C. for a list of U.S. Govt. publications in geology. The following list of textbooks and general references will also help.

Arnold, C. A., AN INTRODUCTION TO PALEOBOTANY, McGraw-Hill, New York, 1947. Background information on fossil plants.

Colbert, E. H., EVOLUTION OF THE VERTEBRATES, John Wiley, New York, 1955. Useful for general reading.

Croneis and Krumbein, DOWN TO EARTH, An Introduction to Geology, Univ. of Chicago Press, Chicago, 1936. A readable, lavishly illustrated account of geology on a college level.

Fenton and Fenton, THE FOSSIL BOOK, Doubleday, Garden City, N.Y., 1958. An introduction to paleontology with fine illustrations.

Matthews, W. H., FOSSILS, Barnes and Noble, New York, 1962. Excellent survey and introduction to prehistoric life.

Moore, Lalicker and Fischer, INVERTEBRATE FOSSILS, McGraw-Hill, New York, 1952. Useful for fossil identification.

Moore, R. C., INTRODUCTION TO HISTORICAL GEOLOGY, McGraw-Hill, New York, 1958. A good source for the geologic history of North America.

Rhodes, F. H. T., THE EVOLUTION OF LIFE, Penguin Books, Baltimore, Md., 1962. An illustrated, readable account of the history of life.

Simpson, Pittendrigh and Tiffany, LIFE: An Introduction to Biology. Harcourt, Brace & Co., New York, 1957. An introductory text on biology. Comprehensive and excellent.

Studying fossils in a museum AMNH

MUSEUMS AND EXHIBITS will show you excellent specimens and introduce you to fossils that do not occur locally. Many universities and most large cities have museums with fossil collections. Some outstanding ones are listed below.

Ala., University—Ala. Museum of Natural History
Ariz., Holbrook — Petrified Forest National Monument Museum
Cal., Los Angeles—L.A. Co. Mus.
Colo., Denver—Mus. of Nat. Hist., Boulder—U. Colo. Mus.
Conn., New Haven—Peabody Mus. of Nat. Hist. (Yale Univ.)
D.C., Washington — Smithsonian Inst., U.S. National Museum
Fla., Gainesville—Fla. State Mus., University of Florida
Ill., Chicago—Chic. Nat. Hist. Mus., Springfield—Ill. State Mus.
Kansas, Lawrence—U. of Kansas Mus. of Natural History
Mass., Cambridge—Museum of Comparative Zoology
Mich., Ann Arbor—University of Michigan Museum
Neb., Lincoln—Univ. of Nebraska State Museum
New York, Albany — N.Y. State Mus., Buffalo—Buffalo Mus. of Sci., N.Y. City—Amer. Museum of Natural History
Ohio, Cleveland—Cleveland Mus. of Natural History
Pennsylvania, Philadelphia—Acad. of Nat. Sci., Pittsburgh—Carnegie Inst. Museum
So. Dakota, Rapid City—Mus. of So. Dakota School of Mines
Texas, Austin —Texas Memorial Museum
Utah, Jensen—Dinosaur Nat. Mon., Vernal—Utah Field Hse. of Nat. Hist.
Wash., Vantage—Ginkgo Petrified Forest State Park Museum
CANADA, Ottawa, Ontario—Nat. Museum of Canada
Toronto, Ontario—Royal Ontario Museum
Montreal, Quebec—Redpath Museum

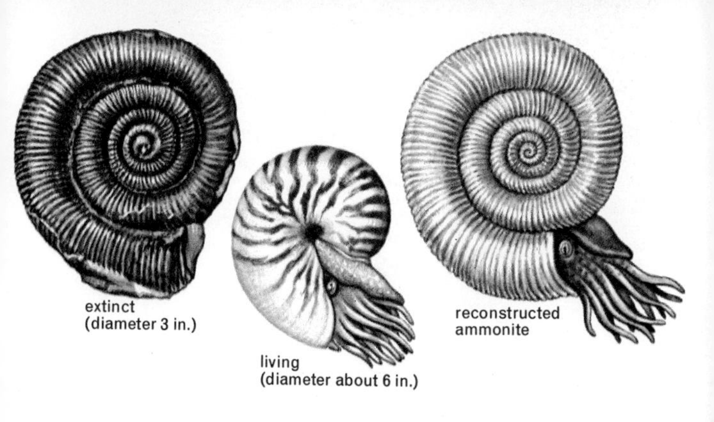
extinct
(diameter 3 in.)

living
(diameter about 6 in.)

reconstructed
ammonite

LIFE OF THE PAST

Fossils are almost always incomplete. A fossil horse is known by a skull and a few bones. Only the shell of a fossil shellfish is found, and an ancient tree is represented only by leaf fragments. Yet entire plants and animals are reconstructed on a scientific basis that uses *present* living forms as a key to interpret the life of the past.

A study of living plants and animals is essential to understand fossils. Fossil ammonites have been extinct for 70 million years, but their shells are very similar to the living Pearly Nautilus. Geologists assume that their soft parts were also similar and make reconstructions accordingly. A comparison of large vertebrate fossils with living species shows how muscles fit to bones. This indicates body structure. Living plants help us understand those known only by fossil fragments. These reconstructions, with other geologic information, make it possible to form an accurate picture of the animal and its environment. This interplay of past and present illustrates how man uses science to develop new frontiers—an increased understanding of both past and present at the same time.

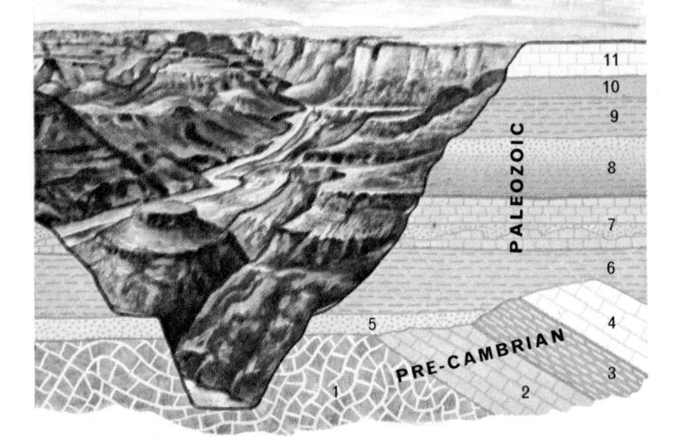

GRAND CANYON rock formations are shown in cross section above, numbered in order of age. The lowest numbers are the oldest. Over 1.5 billion years of earth history are recorded here.

GEOLOGIC TIME is the fourth dimension of the earth's past. Without it, objects and events cannot be placed in their proper relationship. Only the almost incomprehensible length of geologic time can explain the great changes in life and in the earth itself. The development of a reliable scale of geologic time is one of the great feats of the human mind. Early steps were taken by observing the occurrence of sedimentary rocks in horizontal layers and noting the rate at which sediments formed in bays and basins. The simple observation that younger layers formed on top of older ones became the first key to the long geologic time scale.

Studies show that the earth includes a vast series of sedimentary rocks, most with characteristic fossils. Even when layers are tilted, folded and broken, or when erosion has left only discontinuous remnants of strata, fossils reveal their order and relationships.

There are ways by which the age of a rock or fossil can be measured directly in years. One method is based on the breakdown of radioactive elements. These elements have unstable atomic nuclei that break down at a steady, measurable rate to form more stable elements. Thus, uranium breaks down into lead and helium at a very slow rate that is independent of heat, pressure or other conditions. One gram of uranium forms 1/7,000 gram of lead every million years. So a chemist who can accurately measure the ratio of uranium to lead in a rock can get an accurate measure of the age of that rock. When uranium minerals occur in rocks associated with fossils, the age of the fossils can be inferred. This method and others like it, using thorium, rubidium, potassium and carbon, require the most accurate chemical analyses. But, as a result, the geologic time scale is becoming more reliable, year by year.

Other data involving meteorites and the formation of the solar system suggest the earth is four to five billion years old. Fossils two and a half billion years old have been discovered, though fossils did not become abundant until about 600 million years ago.

Geologists know from the rate that sediments form today that much time was needed to make all the sedimentary rocks that total over 75 miles in thickness.

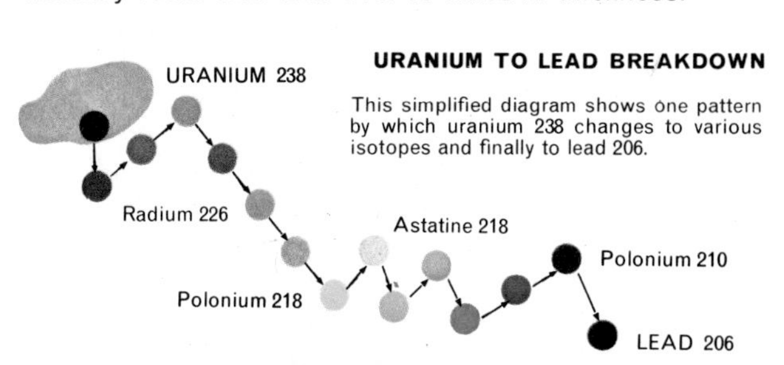

URANIUM TO LEAD BREAKDOWN

This simplified diagram shows one pattern by which uranium 238 changes to various isotopes and finally to lead 206.

Even this time estimate falls short because there were long periods during which sediments were worn away. Yet despite these difficulties the study of unaltered, fossil-bearing sediments shows that they fit into three great eras of time. These eras, in turn, are divided into 12 geologic periods which also have been divided and redivided until each formation can be given a name and a place in the geologic time scale. This record goes back about 600 million years and provides a relative dating for fossils. Yet this time scale (see chart, p. 31) can be and is used every day. We speak of a Jurassic fossil just as we speak of a Colonial mansion and know roughly where both fit into history. Periods are divided and redivided when conditions permit until each strata is identified.

THE GEOLOGIC "CLOCK"

The larger circle represents only the last 600 million years. Each "hour" represents 50 million years.

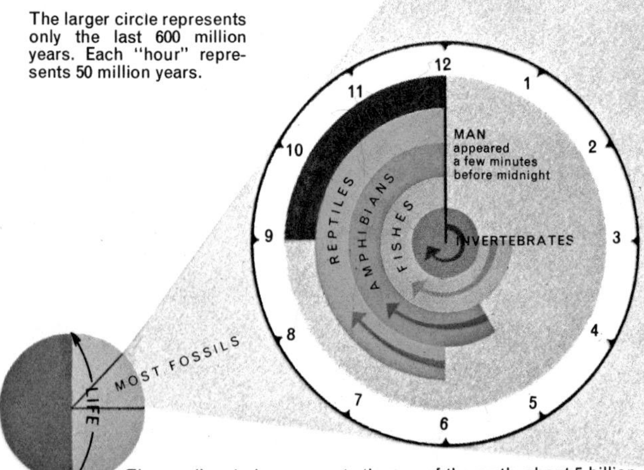

The smaller circle represents the age of the earth, about 5 billion years. Life has existed about half that time and the small segment is 600 million years—the period of abundant fossils.

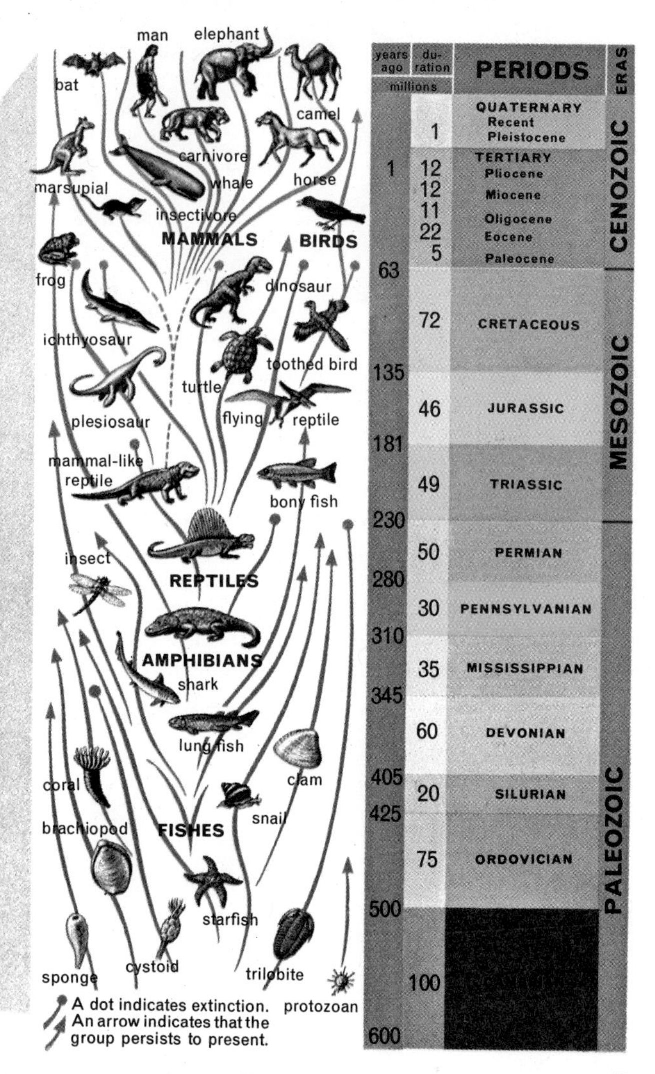
man
elephant
bat
camel
marsupial
carnivore
whale
horse
insectivore
MAMMALS
BIRDS
frog
dinosaur
ichthyosaur
toothed bird
turtle
plesiosaur
flying reptile
mammal-like reptile
bony fish
insect
REPTILES
AMPHIBIANS
shark
lung fish
clam
coral
snail
brachiopod
FISHES
starfish
sponge
cystoid
trilobite
protozoan
A dot indicates extinction.
An arrow indicates that the group persists to present.
years ago
duration
millions
PERIODS
ERAS
QUATERNARY
Recent
Pleistocene
1
TERTIARY
1
12
Pliocene
12
Miocene
11
Oligocene
22
Eocene
5
Paleocene
CENOZOIC
63
72
CRETACEOUS
135
46
JURASSIC
181
49
TRIASSIC
MESOZOIC
230
50
PERMIAN
280
30
PENNSYLVANIAN
310
35
MISSISSIPPIAN
345
60
DEVONIAN
405
20
SILURIAN
425
75
ORDOVICIAN
PALEOZOIC
500
100
600

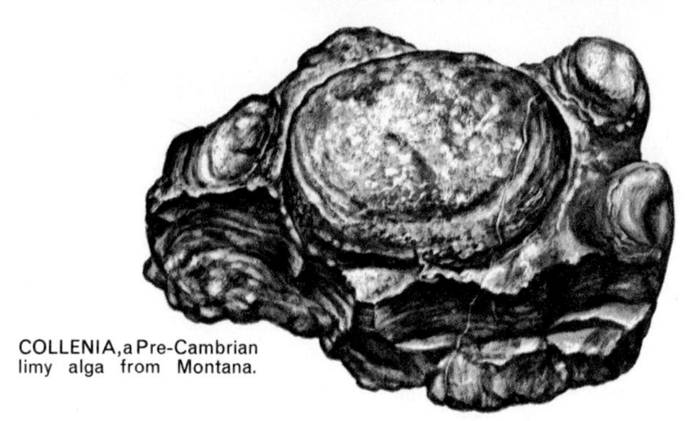
COLLENIA, a Pre-Cambrian limy alga from Montana.

PRE-CAMBRIAN TIME includes the vast period of earth history which elapsed before the deposition of the Cambrian fossil-bearing rocks. It covers a period of about 4,500,000,000 years—or approximately 9/10 of the total age of the earth. This great period of time witnessed the development of the earth, seas and atmosphere, the origin of life, and the early development of living things. But very few fossils of organisms even from the late Pre-Cambrian have been found. Most of those are plants. Lime-secreting algae flourished in the seas of Montana, Alberta and Rhodesia. Pre-Cambrian deposits of anthracite and some

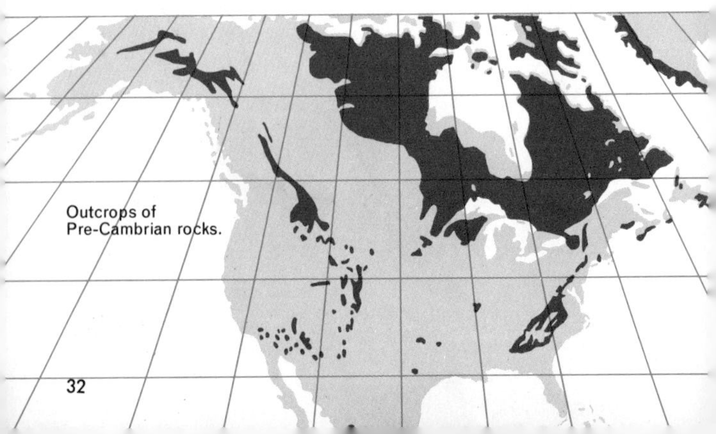
Outcrops of Pre-Cambrian rocks.

limestones are indirect evidence of the existence of life. Primitive aquatic fungi and algae have been found in Pre-Cambrian cherts from Ontario, and in rocks of Michigan, Minnesota, England and Scotland.

Pre-Cambrian animal fossils are rare. A jellyfish is known from the Grand Canyon and some trail-like markings from rocks in Montana. Recently discovered Australian deposits have revealed more animal fossils.

It seems likely that Pre-Cambrian animals were soft-bodied and therefore poorly preserved as fossils. By early Cambrian times, a number of different groups developed hard parts and fossils became more common.

The distribution of Pre-Cambrian rocks is worldwide. They are most extensively exposed in the shield areas, which appear to have remained more or less stable, positive land areas throughout geologic time. The lands of these Pre-Cambrian days must have been startlingly desolate—a barren wilderness of bare rocks. In the shallow seas that lapped these ancient wastelands, life evolved, although fossils give but few clues to the origin of life and its early development. However, biochemical experiments suggest ways that early organic materials may have formed.

Late Pre-Cambrian fossils from Ediacara Hills, South Australia, are the oldest well-preserved animals. (After Glaessner)

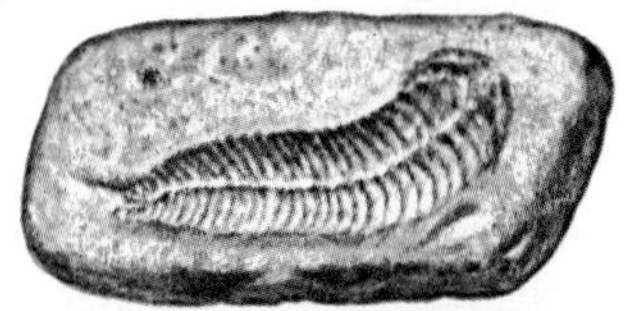

SEGMENTED WORM, *Spriggina floundersi,* 1.5 in.

WORM, *Dickinsonia costata,* 2.5 in.

JELLYFISH, *Medusina mawsoni,* about 1 in.

A Middle Cambrian sea, based on specimens from the Burgess Shale of British Columbia. 1. jellyfish; 2. trilobite *Neolenus* and 3. skeletal remains; 4. arachnid *Sidneyia;* 5. crustacean *Marella;* 6. sponge *Vauxia;* 7. worm *Miskoia;* 8. inarticulate brachiopod *Acrothele*.

PALEOZOIC ERA

THE CAMBRIAN PERIOD (600 million years ago, first period of the Paleozoic) is named from Wales (Latin, *Cambria*), where rocks of this age were first studied. In the Lower Cambrian, the first common and widespread fossils occur: algae, arthropods, brachiopods, sponges, coelenterates, worms, mollusks and echinoderms. All lived in the sea. It is surprising to find so many relatively complex groups in the oldest fossil-bearing rocks. But in many ways Cambrian animals are primitive. Brachiopods are represented by the inarticulate forms (p. 82) and the echinoderms by primitive edrioasteroids. Most trilobites were large, but a few (the agnostids and eodiscids, pp. 94-95) were among the smallest and least ornamented. The first ostracods appeared in the Lower Cambrian. Mollusks were mostly represented by tiny sea snails (gastropods),

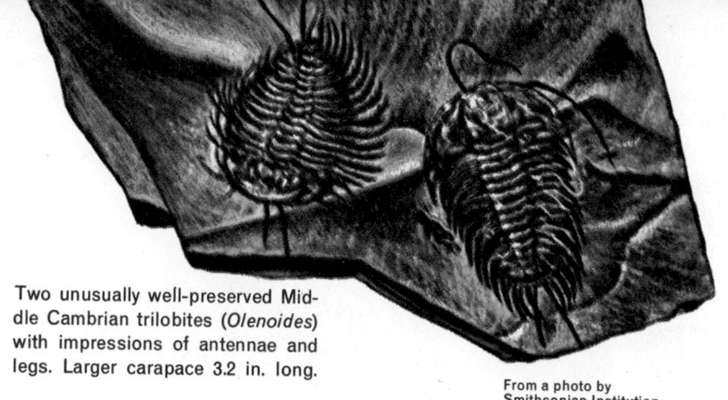

Two unusually well-preserved Middle Cambrian trilobites (*Olenoides*) with impressions of antennae and legs. Larger carapace 3.2 in. long.

From a photo by Smithsonian Institution

but bivalves (pelecypods) appeared in the Upper Cambrian. Cambrian algae were very similar to their very simple Pre-Cambrian ancestors.

Middle Cambrian Burgess Shale in British Columbia contains remarkable fossils—including soft-bodied worms and sea cucumbers. The seas in which these creatures lived occupied two great subsiding belts (geosynclines) in North America. Cambrian rocks have a thickness of over 12,000 ft. in parts of the Rocky Mountains.

See pp. 72-74 and p. 75 onward for more examples and details of the animals and plants shown in the habitat groups of this section.

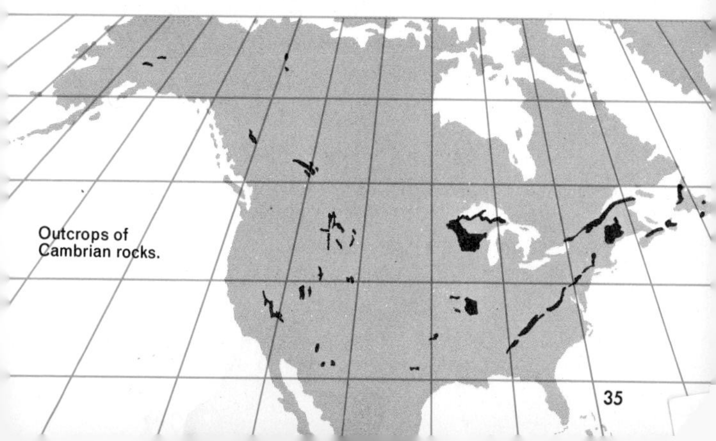

Outcrops of Cambrian rocks.

A Middle Ordovician sea floor, showing straight-shelled, nautiloid cephalopods 1. *Endoceras*, 2. *Sactoceras;* 3. trilobite *Flexicalymene;* brachiopods 4. *Rafinesquina*, 5. *Rhynchotrema;* corals 6. *Streptelasma*, 7. *Favistella;* gastropods 8. *Maclurites*, 9. *Cyclonema;* 10. pelecypod *Byssonychia;* 11. bryozoans *Hallopora*.

THE ORDOVICIAN PERIOD (425 to 500 million years ago) was named for the Ordovices, an ancient Celtic tribe. The Ordovician Period saw the rise of new animal groups of great importance. Bony fragments from the Middle Ordovician of Colorado and Wyoming are evidence of the oldest vertebrates, but we do not yet know much about these fish-like creatures. Tetracorals, graptolites, echinoids, asteroids, crinoids and bryozoans all appeared for the first time, while the articulate brachiopods (p. 82) far outnumbered the inarticulate. Most of the trilobites were different from those of the Cambrian. Some cephalopods reached a length of 13 feet.

In parts of North America and Europe, Ordovician seas covered areas that had been land during Cambrian times. Volcanoes belched lava locally. Uplift and mountain building occurred in eastern North America. Not all the

A fragment of bony armor of an ostracoderm fish (*Astraspis*), the oldest known vertebrate. Middle Ordovician, Colorado.

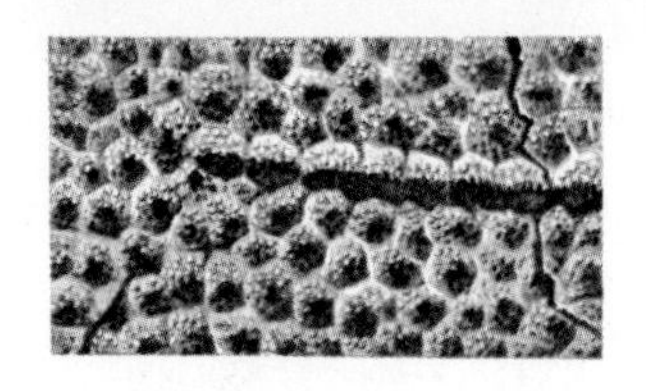

rocks laid down in those ancient seas contain the same fossils. Limestones and shales around Cincinnati, Ohio, contain beautifully preserved brachiopods, corals, bryozoans, mollusks, trilobites and crinoids. Black shales of the same age in New York, Quebec and Wales contain graptolites and occasional trilobites. Different Ordovician environments enabled different animals to prosper in each region. Most common were shallow-water lime and mud deposits noted for their well-preserved fossils.

The repeated and widespread invasion of North America by Ordovician seas has produced extensive Ordovician sediments. Outcrops of these rocks occur widely over much of the continent. Some Ordovician sediments are important oil producers, and Ordovician slates are quarried in Vermont.

Outcrops of Ordovician rocks.

Silurian coral reef showing crinoids 1. *Scyphocrinites,* 2. *Eucalyptocrinites;* 3. starfish; 4. a eurypterid *Eurypterus,* 8 in. long; corals 5. *Favosites,* 6. *Halysites,* 7. *Xylodes;* 8. cephalopod *Cyrtoceras;* 9. gastropod *Beraunia* (*Cyclotropis*); 10. trilobite *Cheirurus.*

THE SILURIAN PERIOD (for Silures, an ancient tribe of the Welsh borderland) lasted from 425 to 405 million years ago. Its faunas differ from those of the Ordovician in the presence of new families and genera, rather than in the appearance of completely new groups of animals. In fact, the most important newcomers are not animals, but plants. Fossils of the oldest land plants come from the Upper Silurian of Australia (p. 151). Fragments of what may be still earlier land plants have recently been found in the Ordovician of Poland and southeastern United States.

Some of the best Silurian fossils (including algae, corals, stromatoporoids, brachiopods, crinoids and trilobites) come from the ancient reefs in Silurian limestones, such as those near Chicago. Scorpion-like eurypterids,

JAMOYTIUS, 10 in., a primitive agnathan fish (p. 132) with scales, lateral eyes, and a terminal mouth. Silurian.

BIRKENIA, an agnathan fish about 4 in. long. Lacks paired fins and true jaws. Silurian, Scotland.

some nine feet long, lived in estuaries and lagoons. Several types of fish are well preserved in parts of the Upper Silurian. Silurian volcanic activity occurred in many areas, and in Scandinavia and Britain mountain building took place at the close of the Period. In other places, deserts and land-locked seas were present in which salt deposits accumulated, such as those of New York, Ohio and Michigan. These deposits are still worked as important commercial sources of salt.

Outcrops of Silurian rocks are common in eastern North America. As in the Ordovician Period, the character of the rocks and of the fossils is evidence of widespread and generally shallow seas.

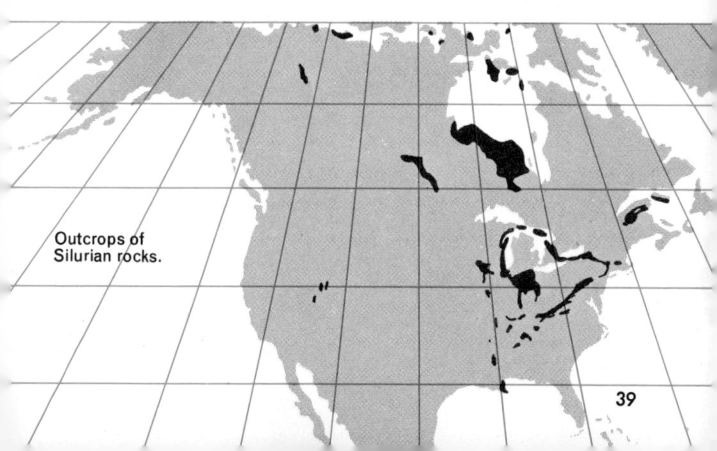

Outcrops of Silurian rocks.

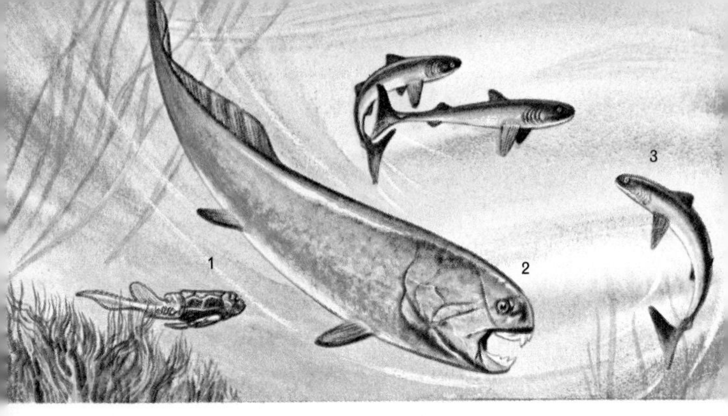

Devonian fishes. A composite diorama showing 1. *Bothriolepis,* a placoderm about 14 in. long, 2. *Dinichthys,* a giant 30-ft. marine placoderm, largest vertebrate of this period, and 3. *Cladoselache,* a 3-ft.-long primitive shark.

THE DEVONIAN PERIOD began about 405 million years ago and ended about 60 million years later. It saw the great expansion of fishes, land plants, and the first land animals, primitive amphibians. The fishes (pp. 132-138) included several kinds of jawless fish (ostracoderms), plate-skinned fish (placoderms), sharks and the first bony fishes (osteichthyes). From one group, the lobe-fin fishes (crossopterygians), the first amphibia (ichthyostegids) arose. These show a mixture of fish and amphibian characters. These unusual fossils, from a warm, moist environment, were found in mountains of Greenland.

Skulls of a Devonian fish, *Eusthenopteron,* and a primitive amphibian, *Ichthyostega.* Corresponding bones are in same color. Note similar bony plates in both skulls but also larger eyes, nostrils, stouter jaws, and loss of gill covering in the head and skull of the amphibian.

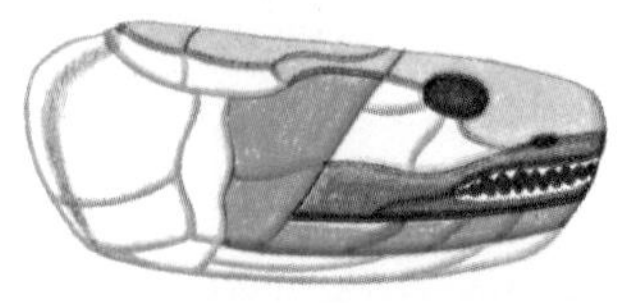

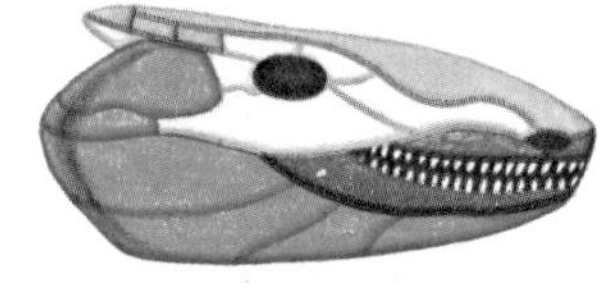

A Devonian forest with 1. a tree tern, *Eospermatopteris;* 2. a small primitive leafless plant, *Psilophyton;* 3. a scouring rush, *Calamophyton;* 4. a primitive lycopod *Protolepidodendron;* and 5. the oldest-known amphibian and first land vertebrate, *Ichthyostega,* 2½ ft. long.

The oldest spiders, millipedes and insects appear in the Devonian, as do fresh-water clams. Early land plants were simple, lacking true roots and leaves, but with the vascular or conducting system found in all later land plants. Late in the Devonian, great forests of scale trees and seed ferns were widespread.

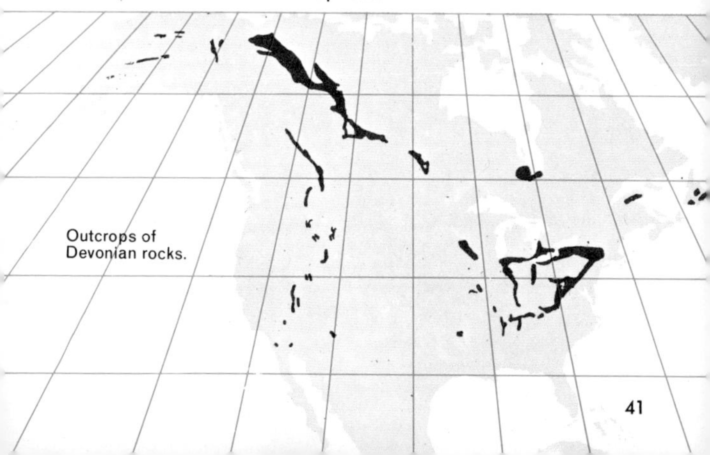

Outcrops of Devonian rocks.

A Devonian coral reef. Corals 1. *Heliophyllum*, 2. *Cylindrophyllum*, 3. *Hexagonaria*, 4. *Synaptophyllum*, 5. *Heterophrentis*, 6. *Pleurodictyum*, 7. *Chonophyllum*; 8. bryozoan *Fenestrellina*; brachiopods 9. *Leptaena*, 10. *Atrypa*; 11. gastropod *Platyceras*; 12. cephalopod *Michelinoceras*; trilobites 13. *Calymene*, 14. *Anchiopsis*; 15. crinoid *Dolatocrinus*.

Devonian coral reefs include large cup corals two feet high and compound corals eight feet across. Horn corals were numerous and varied. Brachiopods and mollusks continued to flourish; the first common ammonites appeared, but true graptolites were already extinct and trilobites were greatly reduced in numbers. In many continental areas thick deposits of red sands and muds accumulated.

A colony of Mississippian crinoids 1. *Cyathocrinites*, 2. *Taxocrinus*, 3. *Batocrinus*, 4. *Barycrinus*, 5. *Scytalocrinus*; 6. a brittle star *Onychaster*.

THE MISSISSIPPIAN PERIOD is named for the limestone bluffs along the Mississippi River where typical outcrops occur. It was a period (345 to 310 million years ago) of shallow, warm seas, in which corals, brachiopods, crinoids, blastoids, bryozoans and foraminifera flourished. In places these fossils are so abundant that they make up most of the rocks. On land, amphibia continued to develop, while land plants spread in all moist areas and anticipated the great coal swamp forests of the Pennsylvanian. Much of North America except the far west and the east coast was under water during Mississippian time.

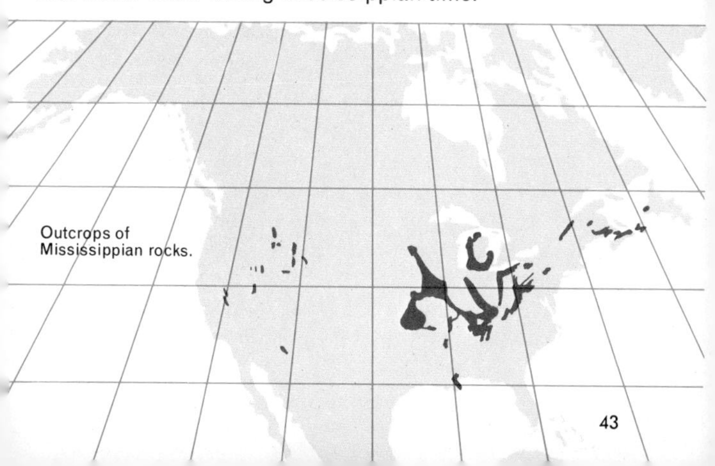

Outcrops of Mississippian rocks.

A Pennsylvanian coal swamp. Trees include lycopods 1. *Sigillaria,* 2. *Lepidodendron;* 3. sphenopsid *Calamites;* 4. seed ferns; 5. *Cordaites;* 6. a labyrinthodont amphibian; 7. *Meganeura,* an insect with a 30-in. wingspan related to the modern dragonflies.

THE PENNSYLVANIAN PERIOD (310 to 280 million years ago) was named after the great coal-bearing strata of Pennsylvania. It saw the development of lowlands, great swamps, and deltas surrounded and often covered by shallow seas. Some of the land was barren sand deserts (English Midlands), or salt basins (Colorado). Great trees, some 150 feet high, formed the coal forests in low swampy land that was often flooded. Most common were the scale trees (lycopods), seed ferns (pteridosperms), horsetails and cordaites. Here lived giant "dragonflies," with a 30-

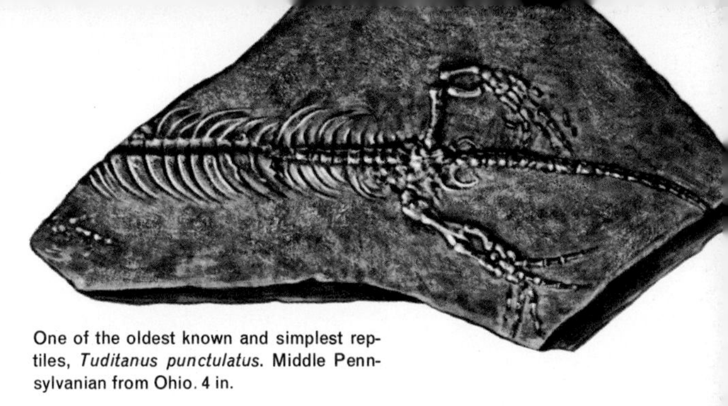

One of the oldest known and simplest reptiles, *Tuditanus punctulatus*. Middle Pennsylvanian from Ohio. 4 in.

inch wingspan, and many kinds of amphibians. Pennsylvanian rivers and deltas were inhabited by countless clams, other shellfish and fishes. This period also saw the emergence of the reptiles from amphibian ancestors. In addition to *Tuditanus* (above), fossils representing four groups of primitive reptiles have been found in the shales of Kansas. The seas continued to support rich invertebrate life, which included abundant spindle-shaped foraminifera (fusulinids), corals, brachiopods, mollusks, bryozoans, crinoids, ostracods and a few trilobites.

Outcrops of Pennsylvanian rocks.

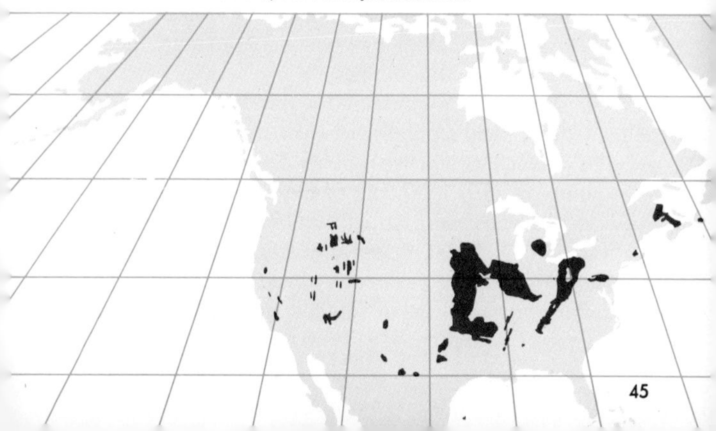

A Permian landscape. Reptiles include 1. *Dimetrodon,* 10 ft. long, a "sail-backed" carnivore; 2. *Seymouria,* 2 ft. long. Amphibia include 3. *Eryops,* 6 ft. long and 4. *Diplocaulus,* 2 ft. long.

THE PERMIAN PERIOD (280 to 230 million years ago) began with typical coal-forest plants, which were later replaced by primitive conifers, especially in semi-arid upland regions. In parts of the Southern Hemisphere the most common plants were a distinctive group of tongue ferns (*Glossopteris*). Many new insects appeared, including beetles and true dragonflies.

Streams and ponds contained a variety of fishes. Amphibians flourished along their banks, but were overshadowed by newer, more active reptiles. Early reptiles differed from amphibia only in details of the skull and vertebrae. Seymouriamorphs were squat, lumbering reptiles about two feet long, with flat, massive heads. Fossil eggs from the Lower Permian of Texas, the oldest land eggs known, may belong to them. Other reptiles were quite different. *Dimetrodon,* the sail-backed lizard, was a savage carni-

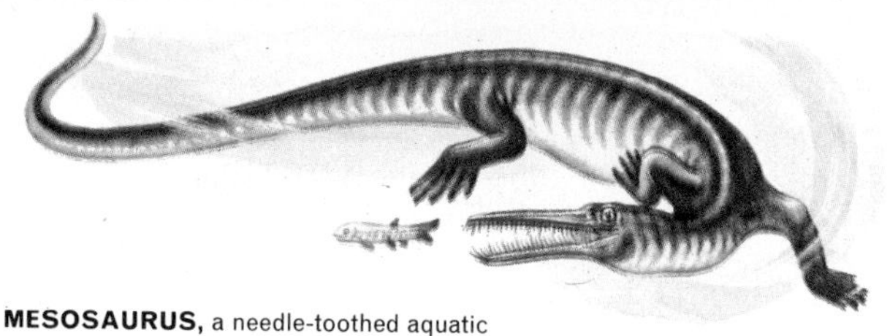

MESOSAURUS, a needle-toothed aquatic reptile. Pennsylvanian-Permian. 16 in.

vore, about 10 feet long. *Edaphosaurus,* a vegetarian, was also a sailback. The purpose of these sails is obscure. They may have served as primitive temperature controls.

Other Permian reptiles included mesosaurs, small, long-snouted, aquatic creatures, and other species similar, but unrelated, to modern lizards. Another group, the theriodonts (beast teeth), known from South Africa and Russia, were small, agile carnivores, from which mammals are descended. *Cynognathus* was a typical theriodont, about 6 feet long, with a doglike skull, and differentiated teeth. Its legs, placed below the body, lifted it clear of the ground. This was a better adaptation to a

A Permian reef in west Texas. Brachiopods 1. *Dictyoclostus,* 2. *Dielasma;* 3. productid; 4. *Leptodus* shells; sponges 5. *Girtyocoelia,* 6. *Heliospongia;* cephalopods 7. *Stenopoceras,* 8. *Cooperoceras.*

more active life than the sprawling legs of amphibians and primitive reptiles.

The close of the Permian marked the end of the Paleozoic Era—the first great chapter in the recorded history of life. By then, many animals and plants which had dominated the Paleozoic scene had become extinct. Fusulinid foraminifera, various bryozoans, rugose corals, productid brachiopods, trilobites and blastoids all vanished, as well as many crinoids and cephalopods. Giant scale trees dwindled in numbers. Most horsetails and many ferns became extinct. Amphibia and some fish underwent a drastic reduction. Why this happened is not clear, but it may have been connected with extreme climatic changes during the late Permian when seas were very restricted and large, high continents emerged. In many areas, coral reefs fringed the shores of deserts and vast inland salt lakes formed. Extensive glaciers covered parts of the Southern Hemisphere. New mountain chains slowly rose, the Appalachians and the Urals among them.

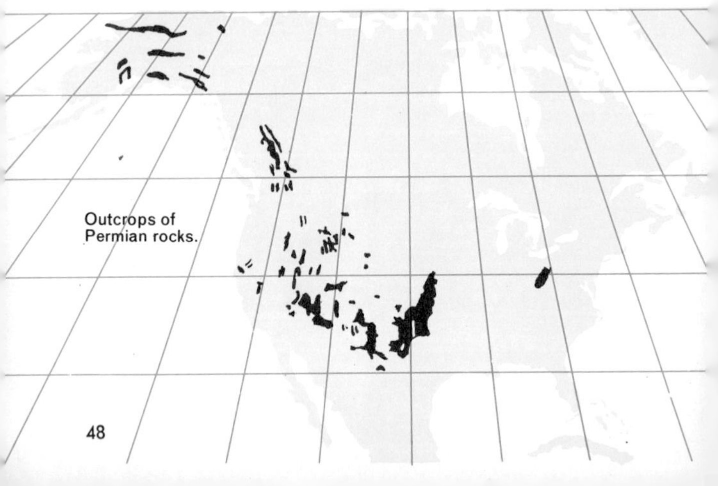

Outcrops of Permian rocks.

TYRANNOSAURUS, the largest carnivorous dinosaur, was about 50 ft. long with a 4- to 5-ft. skull. Cretaceous of Montana.

THE MESOZOIC ERA

The Mesozoic (Middle Life) Era covers a period of about 165 million years, during which reptiles so overshadowed all other animals that it is often called the "Age of Reptiles." Great changes took place in some invertebrates too. New forms replaced those which had become extinct at the end of the Paleozoic. Ammonites developed rapidly until countless numbers lived in the seas. Birds, mammals, flowering plants and many modern insects appeared for the first time. Elm, oak, maple and other modern broad-leaved trees became common. The development and spread of some flowering plants depended on the parallel development of insects which pollinated the flowers.

Other important geographic changes were also taking place. New patterns of lands and seas formed. New mountain ranges slowly emerged. As the result of several related geologic processes, great mineral deposits were formed. Sixty million years later, we still depend on many of these deposits for our metals and our fuels.

A Triassic semi-arid landscape. The reptiles include 1. *Cynognathus*, a 7-ft. mammal-like carnivore; 2. *Machaeroprosopus*, an alligator-like phytosaur; 3. *Saltoposuchus*, a 4-ft. thecodont; 4. *Kannemeyeria*, a 6-ft. herbivorous dicynodont more common in the uplands.

THE TRIASSIC PERIOD (230 to 180 million years ago) was named from a threefold division of its rocks. In many places Triassic rocks resemble those of the Permian Period—thick sequences of red shales and sandstones, deposited in temporary lakes, deserts and basins. Volcanic activity was considerable, as in eastern North America from Virginia to Connecticut.

Against this background the reptiles developed and established their mastery. Their advanced body structure and shell-protected eggs enabled them to survive changing and often adverse climates, and to colonize new areas which were forbidden to the water-tied amphibia. The first dinosaurs appeared; their footprints are abundant in some rocks, as in the Connecticut Valley.

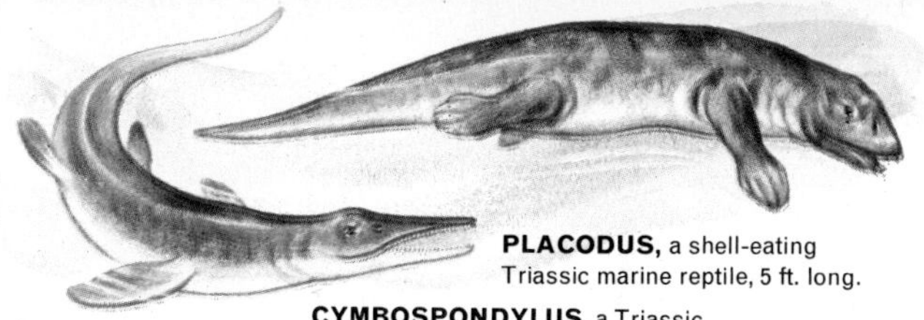

PLACODUS, a shell-eating Triassic marine reptile, 5 ft. long.

CYMBOSPONDYLUS, a Triassic marine reptile, 23 ft. long.

Nor was the dominance of reptiles confined to land, for in open oceans dolphin-like ichthyosaurs swept through the water. Later 15- to 20-foot long plesiosaurs paddled their way through Triassic seas.

New types of sponges and protozoans developed. The modern hexacorals appeared, and new groups of brachiopods replaced their Permian forebears. Gastropods and pelecypods increased in numbers. Ammonites flourished and underwent considerable change. Lobster-like arthropods and modern echinoids and crinoids first appeared in the Triassic.

Cycads and primitive conifers flourished on upland areas. The Petrified Forest of Arizona contains fossils of these trees. Ferns and scouring rushes prospered in lower, moist areas.

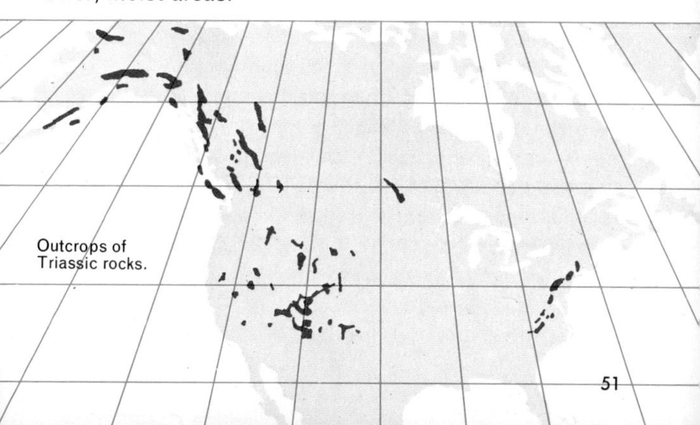

Outcrops of Triassic rocks.

Generalized diorama of Jurassic marine reptiles showing 1. *Ichthyosaurus,* 10 ft. long; 2. *Plesiosaurus,* about 12 to 20 ft. long; 3. *Eurhinosau-*

THE JURASSIC PERIOD is named from the Jura Mountains. It began about 180 million years ago and lasted about 45 million years. Of all its abundant and exotic life, none was more typical than the dinosaurs of which there were three main Jurassic groups: first, the sauropods, long-necked, long-tailed, four-legged monsters, which included the largest land animal (*Diplodocus,* 87 ft. long); second, stegosaurs, armored reptiles that weighed up to 10 tons (with only a 3-oz. brain); third, the carnivorous theropods which walked on their hind legs, including *Allosaurus,* a savage, 35-ft. creature. Others were more slender, and some were only 3 feet long. A few early,

rus, about 17 ft. long; 4. *Cryptocleidus,* about 10 ft. long, Upper Jurassic; 5. squid-like belemnites fleeing in a protective cloud of "ink."

duck-billed herbivorous dinosaurs lumbered across the swampy lowlands.

Flying reptiles gliding through the air included sparrow-sized species and others up to 4 feet long with slender club-like tails. Ichthyosaurs and plesiosaurs were the carnivorous masters of the oceans. Hosts of ammonites (some up to 6 feet in diameter) thronged the shallow seas, together with gastropods, pelecypods, squids (belemnites), echinoids, crinoids and foraminifera.

The Jurassic also saw the development of two groups that were later to establish their dominance. The oldest mammals are known from fossil fragments of rat-sized

Shore of a Jurassic lagoon at Solenhofen, Germany. 1. first bird, *Archaeopteryx;* 2. flying reptiles, *Rhamphorhynchus;* 3. small bipedal dinosaur, *Compsognathus* about 2 ft. long; and 4. cycadeoid plants.

jaws and teeth from western United States and Europe. The Solenhofen Jurassic limestone of Bavaria contains remains of *Archaeopteryx,* the oldest known bird.

Jurassic plants included the now extinct cycadeoids with short, thick trunks. These were crowned with frond-like leaves and ornate reproductive structures which closely resembled modern flowers. Cycads, conifers, ferns and ginkgos were common. Ginkgos were widespread all through the Mesozoic but later became almost extinct. Today only a single species survives but this is widely planted. Over a thousand species of insects are known, including many modern forms.

A Jurassic swamp with herbivorous dinosaurs. 1. *Brontosaurus*, a 67-ft. long sauropod; 2. *Diplodocus*, a similar, but more slender, 87-ft.-long animal; 3. *Stegosaurus*, a 20-ft.-long armored dinosaur.

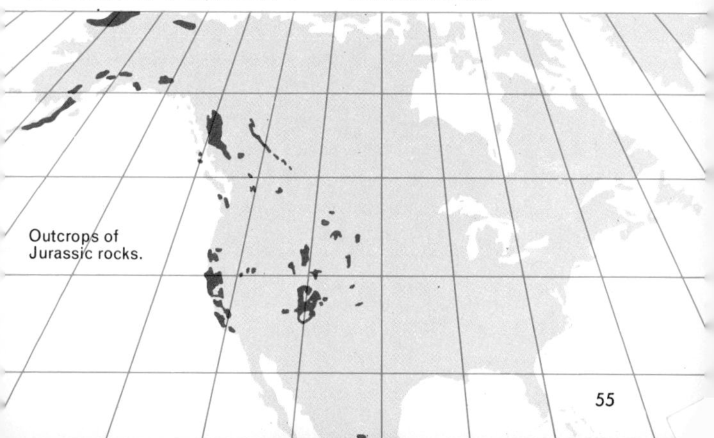

Outcrops of Jurassic rocks.

A diorama of Cretaceous reptiles. Dinosaurs include 1. *Triceratops*, young and adult, 20 ft. long; 2. *Trachodon*, a 40-ft. duck-bill dinosaur; 3. *Tyrannosaurus*, a powerful carnivore, 50 ft. long; 4. ostrich-like *Stru-*

THE CRETACEOUS PERIOD (named from chalk, its most characteristic deposit) began about 135 million years ago and lasted some 70 million years. It was one of the most important of all geologic periods, marked by a major advance of the sea in many parts of the world, and by the great thickness of both marine and continental sediments. A great depression connected the Arctic Ocean with the Gulf of Mexico. Middle Europe was also submerged except for a central land mass. Towards the close of this period earth movements produced mountain ranges which are now the Andes and the Rockies, as well as mountains in Antarctica and northeastern Asia.

The Cretaceous Period marked both the culmination of Mesozoic life and the foreshadowing of animals and

thiomimus, 5 ft. high; 5. *Brachiosaurus* (Lower Cretaceous); 6. *Pteranodon,* a flying reptile with a 25-ft. wingspan. Plants include 7. cycadeoids and 8. primitive angiosperms or flowering plants.

plants that were later to displace it. The most important new arrivals were the flowering plants (angiosperms). They first appeared in the Lower Cretaceous, but eventually they became the dominant plants on every continent. Many familiar living trees and shrubs, including the poplar, magnolia, oak, maple; beech, holly, ivy and laurel appeared during the Cretaceous. The spread of the flowering plants also had important effects on animal life, for they provided new sources of food for mammals, birds, reptiles and insects. The subsequent expansion of mammals and birds depended very largely upon these new food supplies.

Dinosaurs extended their dominance across Cretaceous lands. They are known from every continent, and included

In a Cretaceous seascape 1. *Tylosaurus,* a 26-ft. mosasaur, chases 2. *Archelon,* a 12-ft. marine turtle. Hammerheaded flying reptiles, 3. *Pteranodon,* with a 25 ft. wingspan, are overhead.

many unusual types. Horned dinosaurs (ceratopsians) were common, as were the armored ankylosaurs, and the bizarre duck-bills with their striking amphibious adaptations. The great quadruped dinosaurs declined in the Cretaceous but savage carnivores were common. *Tyrannosaurus* stood 20 feet high and had a skull over 3 feet long. Other carnivores were much smaller. Flying reptiles were represented by *Pteranodon,* a toothless, hammerheaded creature with a wingspan of 25 feet, the largest animal ever to fly.

In the seas, giant turtles (*Archelon*) reached a length of 12 feet, and some plesiosaurs grew over 40 feet long. Ichthyosaurs declined. Savage, serpent-like mosasaurs, some 35 feet long, were sea-going lizards.

A late Cretaceous sea floor showing a rich variety of mollusks. Ammonites 1. *Helioceras*, 2. *Baculites*, 3. *Placenticeras;* gastropods 4. *Turritella*, 5. other; pelecypods 6. oysters, 7. *Pecten*.

Two well-known types of fossil birds occur in Cretaceous rocks. *Ichthyornis,* a slender, tern-like bird about 8 inches high, was a strong flier. *Hesperornis,* in complete contrast, was about 4½ feet high, a diving bird, with powerful swimming legs but only vestiges of wings. It also had long, toothed jaws.

Mammals were small and relatively insignificant. Their remains are rare, represented by small primitive forms that survived from the Jurassic and also by two new groups, the opossum-like pouched marsupials and the insectivores, forerunners of the shrews. The fossils are mostly teeth and parts of lower jaws, which, because of their unique structures, are sufficient to distinguish these true mammals from mammal-like reptiles.

In the shallow seas invertebrates lived in great diversity. The dominant group was the ammonites, which showed many unusual forms. Belemnites, pelecypods and gastropods of rather modern appearance, corals, sea urchins and foraminifera also flourished. Modern bony fish (teleosts) were common. The corals, abundant locally in Cretaceous beds, show a basic sixfold symmetry. Crinoids also developed new forms, including a free-swimming, stemless crinoid with slender arms up to 4 ft. long. *Inoceramus* mollusks (p. 118), sometimes 3 to 5 ft. across, were widely distributed.

The close of the Cretaceous saw the widespread extinction of many of the dominant animals of the Mesozoic Era. Dinosaurs, pterosaurs, ichthyosaurs, plesiosaurs, mosasaurs, ammonites, true belemnites, many pelecypods and corals all became extinct. It was as truly the end of an era in the long history of life as was the Permian, and the causes for this widespread decline are no less difficult to identify. It is probable, however, that the great geological changes, and the changes in plants exercised a profound effect on many groups of animals.

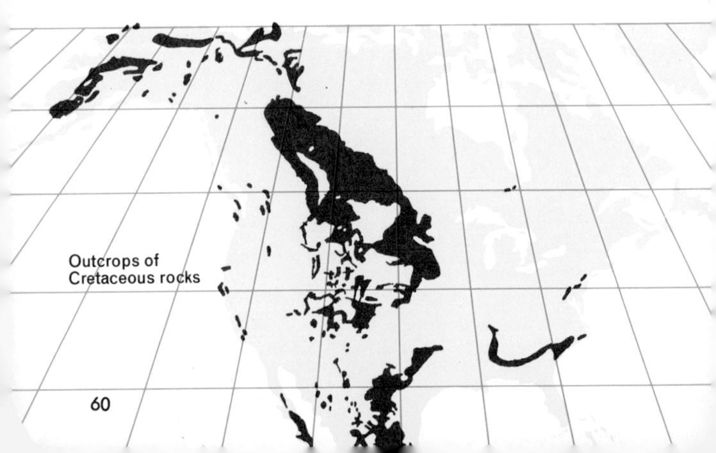

BRONTOTHERIUM, an 8-ft.-high North American grazing mammal, one of the extinct titanotheres. Oligocene of South Dakota.

THE CENOZOIC ERA

The Cenozoic Era includes the last 70 million years of earth history. It is far more familiar than previous eras, for although some animals and plants died out other kinds have survived without drastic change and are alive today. Slow but striking changes in climate took place during this time. Polar regions cooled and the general warm temperate climate gave way to a wider climatic range. The continents were similar to those of today, although there was mountain building, continental warping and volcanic activity. Cenozoic strata in the Gulf Coast, California, the Middle East and the East Indies are now important petroleum producers. Cenozoic life (the Age of Mammals) is dominated by the mammals and flowering plants. Mammals replaced the ruling reptiles of the Mesozoic in every environment. The flowering plants became broadly similar to living forms. Amphibia and reptiles became relatively inconspicuous. Birds continued to expand in numbers and variety. The bony fish (teleosts) outnumbered all other fish by twenty to one.

Marine invertebrates took on a modern look. Gastropods and pelecypods became the most abundant, and cephalopods and brachiopods were greatly reduced. This then is the last great era in the long history of life—the prelude to the present.

A diorama of Lower Tertiary life. 1. *Diatryma*, a 7-ft.-high carnivorous bird. Mammals include 2. *Notharctus*, a lemur-like primate; 3. *Coryphodon*, a 3-ft.-high amblypod; 4. *Hyracotherium* (Eohippus) the ancestral 4-

THE LOWER TERTIARY PERIOD (Paleocene, Eocene, and Oligocene) in North America is represented by great continental, badland deposits which contain many mammalian fossils. Thick marine deposits formed in south-eastern U.S. and along the Pacific Coast where volcanoes also erupted.

Lower Tertiary fossils are strikingly different from those of the Cretaceous. Mammals increased explosively, spreading into all environments and becoming adapted to many ways of life on land, water and in the air. Mammalian history varied. In isolated Tertiary South America a great diversity of marsupial mammals developed. Other marsupials still survive in the isolation of Australia.

Early Tertiary mammals included primitive rodent-sized forms, insectivores and marsupials, that had persisted

toed, 12-inch-high horse; 5. *Uintatherium,* a rhinoceros-sized horned mammal. Trees are palms and broad-leaved angiosperms similar to modern temperate or semi-tropical species.

from the Cretaceous. New species included two main groups, hoofed mammals and carnivores.

Creodonts, forerunners of the carnivores, and condylarths, forerunners of hoofed mammals, were somewhat similar, differing mainly in details of teeth and feet. Both were squat, heavy-limbed, about the size of a collie, with dog-like heads and five bluntly clawed toes on each limb. Their brains were small and primitive. Other new arrivals were ancestral rodents and primates, represented by lemur-like animals.

Amblypods were heavy, clumsy, hoofed mammals, with broad feet. The early ones were sheep-sized, but the later kinds (uintatheres) were seven feet high, as big as a large rhinoceros, with three pairs of blunt horns. The males had large down-curved tusks.

An Oligocene landscape includes 1. *Mesohippus*, a three-toed horse, 2 ft. high; 2. *Brontops*, a 14 ft. horned titanothere; 3. *Oreodon*, a sheep-sized herbivore; 4. *Baluchitherium*, a 18-ft.-high rhinoceros, the largest land mammal; 5. *Protapiris*, a tapir; 6. *Hyaenodon;* a creodont carnivore; and, 7. land tortoises. *Stylemys*, (feeding on grasses).

During Eocene times more advanced mammals displaced many older forms in North America and Europe. These new arrivals included true rodents, mainly squirrel-like species, and a variety of rhinoceroses (one Oligocene giant, *Baluchitherium*, measured 18 feet high at the shoulders). Ancestral tapirs, titanotheres and the first even-toed hoofed mammals also appeared. The early titanotheres were small, hornless browsers with clumsy bodies and tiny brains. Condylarths, creodonts and uintatheres persisted for a time, but lost out to later arrivals such as early horses (the three-toed, collie-sized *Mesohippus*), giant pigs, ancestral camels, oreodonts, mastodons and large saber-toothed cats.

Oreodonts were sheep-sized, long-tailed herbivores that survived until the Pliocene. Titanotheres were huge, grotesque horned beasts. The cat family first appeared in the form of the earliest saber-tooth, *Hoplophoneus,* about the size of a mountain lion.

Lower Tertiary birds were modern in appearance, but included numbers of large, ground-living genera, one of which, *Diatryma,* was 7 feet high. Fish, too, were of modern aspect. The fine fresh-water sediments of the Green River beds of Wyoming have yielded thousands of beautiful, well-preserved fish fossils.

Marine invertebrates were very much like modern forms. Large foraminifera ("*Nummulites*") abounded in the shallow seas of the Mediterranean and Caribbean. Plants resembled living forms, but palms grew in Canada, and temperate oak and walnut forests in Alaska. There was mountain building and crustal disturbance in the Alps, Carpathians, Pyrenees, Apennines, and Himalayas. The Coast Range of western North America was the scene of mountain building too, with volcanic activity.

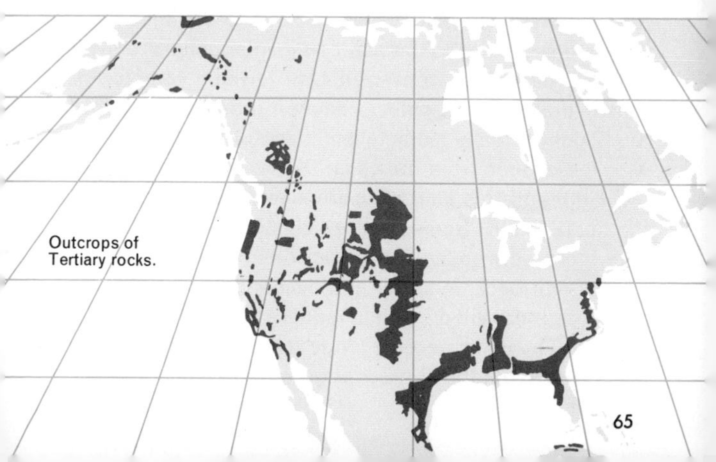
Outcrops of Tertiary rocks.

A Miocene landscape. 1. a wild pig, *Dinohyus;* 2. a small rhinoceros, *Diceratherium;* 3. horse-like *Moropus;* 4. a four-tusked mastodon, *Trilophodon;* 5. a herd of primitive camels, *Stenomylus.*

THE UPPER TERTIARY PERIOD (Miocene and Pliocene) lasted about 25 million years, ending about a million years ago. It is marked by the continued rise of modern mammals. Changes involving brain, limbs and teeth and the accompanying expansion of mammals as a group were closely related to climatic changes. Over wide areas of North America continental uplift produced drier climates, and converted lush lowland forests into grassy prairies. The oldest common grasses come from the Miocene. Many mammalian changes were associated with the change from browsing to grazing habits.

Changes of this kind are particularly well illustrated in the horse family. Some of these changes were correlated with increase in overall size, but others were not. Thus horse teeth became larger and deeper, but they also became high-crowned, with a square, infolded, chewing surface. The limbs became longer and changed in relative proportions, but the number of ground-touching toes was

EVOLUTION OF THE HORSE

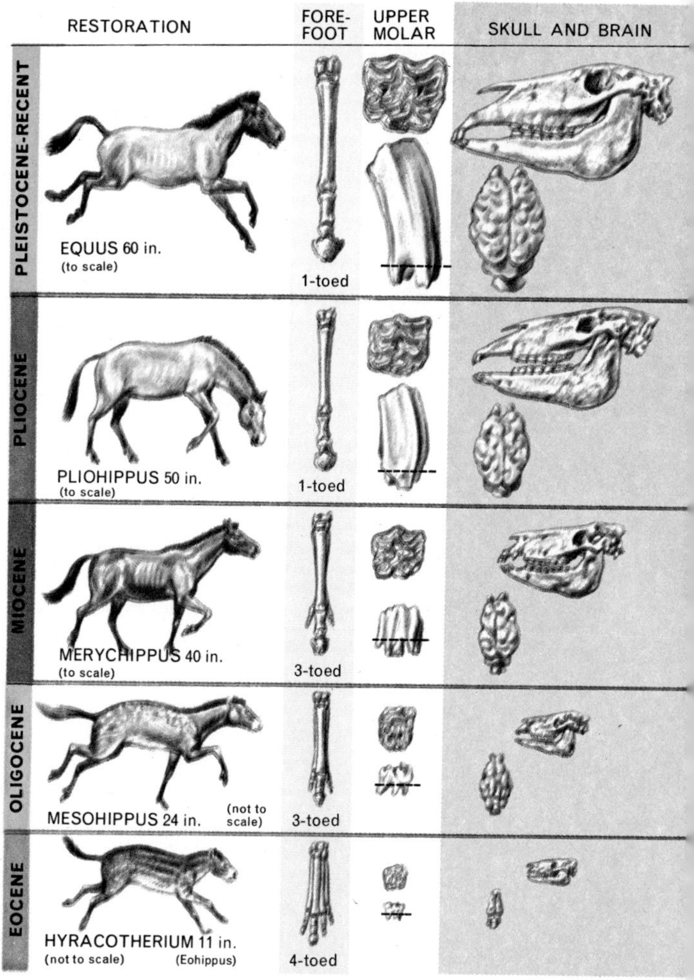

	RESTORATION	FORE-FOOT	UPPER MOLAR	SKULL AND BRAIN
PLEISTOCENE-RECENT	EQUUS 60 in. (to scale)	1-toed		
PLIOCENE	PLIOHIPPUS 50 in. (to scale)	1-toed		
MIOCENE	MERYCHIPPUS 40 in. (to scale)	3-toed		
OLIGOCENE	MESOHIPPUS 24 in. (not to scale)	3-toed		
EOCENE	HYRACOTHERIUM 11 in. (not to scale) (Eohippus)	4-toed		

reduced. This reflects a radical change from a flat-footed to a tip-toe, spring-hoofed posture. Feeding on tough prairie grasses demanded tougher teeth, and the advantage of speed on the hard, open prairies favored the new foot structure.

Widespread changes also took place in other groups of mammals. Ancestral elephants, camels, rhinoceroses, dogs and smaller carnivores abounded. Other forms have no living descendants. *Moropus* resembled a clumsily constructed horse with claws. One giant pig had a skull four feet long. There were giraffes, camels, and the pathetic antelope, *Syndyoceras,* with the strangest horns of any of its tribe. Saber-toothed cats continued, and apelike creatures (*Dryopithecus*) spread across Europe and Africa. Many of the older type mammals became extinct towards the close of Pliocene times.

Most Upper Tertiary marine invertebrates and plants are barely distinguishable from modern species. Renewed earth movements in the Alps, Himalayas and along the Pacific Coast of North America complicated and extended existing mountain ranges.

Our hurried excursion through geologic time has taken us, with giant steps, over a period of more than half a billion years from the late Pre-Cambrian to the end of the Tertiary. The late Tertiary world is modern except for minor species development including the explosive dominance of man. Though fossil clues are meager, the panorama of life is understandable because of the length of geologic time. The slow organic changes which led to extinction or survival are the building blocks of evolution and, for their operation, time is needed. Man's development is a matter of only a few million years, but a billion years of preparation lie behind it.

A Pleistocene landscape with 1. woolly mammoth, *Elephas primigenius*, 12 ft. high; and, 2. a woolly rhinoceros, *Coelodonta*, 6 ft. high.

THE QUATERNARY PERIOD includes the moment in which we now live, together with the Pleistocene Epoch, which began about one million years ago. Continental ice sheets, up to ten thousand feet thick, spread over much of the Northern Hemisphere in at least four glacial advances, the last of which retreated about 11,000 years ago in North America. Antarctica and the mountains of the Southern Hemisphere were also glaciated.

There is evidence of repeated floral and faunal migrations in response to climatic changes. During the colder episodes vast herds of wild pigs, camels, bison and elephants ranged across North America, Europe and Asia. There were four American species of elephants, including the Imperial Mammoth, 14 feet high at the shoulder, with curved tusks 13 feet long.

IRISH ELK (*Megaceros*), Pleistocene, Europe, had antlers up to 11 ft. across, the largest known deer.

SMILODON, largest of saber-toothed cats, had upper canine teeth 8 in. long. North America. Pleistocene.

Majestic woolly mammoths, which roamed across the tundra of Europe, Asia and North America, are pictured in early cave paintings. Most of these large mammals became extinct near the end of the Pleistocene. Carnivores, including wolves, foxes, badgers, and the terrible sabertooth *Smilodon,* are well known from the Pleistocene tar pools of California. Huge armadillo-like glyptodonts and giant (20 ft. high) ground sloths had evolved in South America and spread into North America when the land bridge between the two continents was re-established in late Pliocene times. In this arid and partly frozen world, man emerged.

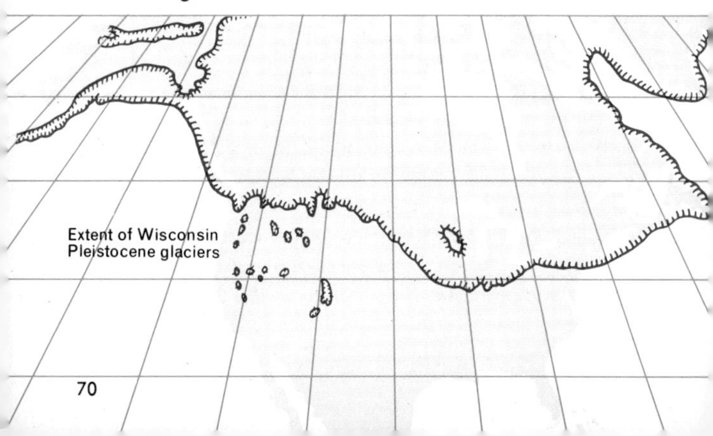

Extent of Wisconsin Pleistocene glaciers

THE EMERGENCE OF MAN

CRO-MAGNON man, finely built, tall, muscular, with modern brain and facial features, replaced Neanderthal. He manufactured finely worked tools from stone, ivory and bone, practiced ceremonial burial and was probably advanced socially as well. Cro-Magnon cave paintings, drawings and sculpture are beautifully executed.

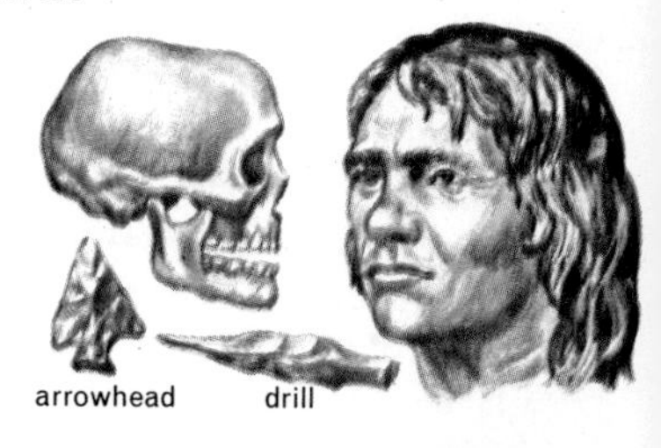

NEANDERTHAL man lived over a wide area of Europe and North Africa during the last glacial advance. He was short, stocky, stooped, with heavy brows, retreating forehead and a prominent but chinless jaw. A cave dweller and skilled hunter, he practiced ceremonial burial. He became extinct after about 100,000 years. Not ancestral to modern man, but probably from same stock.

PITHECANTHROPUS (the ape man) is known from specimens about 500,000 years old from Java and China. Height about 5 ft., semi-erect posture, heavy brows, powerful jaws with manlike teeth and a brain capacity between that of the large apes and modern man. Charred bones found with simple stone tools suggest cannibalism.

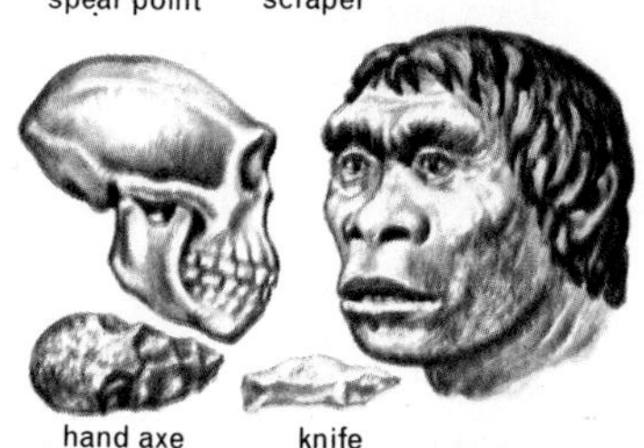

AUSTRALOPITHECUS of So. Africa and his relative, *Zinjanthropus,* of Olduvai Gorge, E. Africa, are more than 1½ million yrs. old. A recent Olduvai discovery is an advanced fossil man, *Homo habilis,* a hunter who probably built simple shelters and made most of the tools found with the split bones of game he ate. Other discoveries have pushed the age of man's direct ancestors back to about 20 million years.

INVERTEBRATE FOSSILS

The fossils described on pages 72-129 all represent animals that lack backbones (invertebrates). Most lived in the sea, and most are now extinct. For these reasons they are less familiar than most vertebrates and plants, and the terms used to describe their structure are not widely known. However, small structures are important in identification of invertebrates. The following diagrams (pages 72, 73, 74) show structures of the most common invertebrate fossils. They provide a guide to ten major groups, and indicate the meanings of terms used later in the text.

For more about invertebrate fossils read:

Buchsbaum, R. ANIMALS WITHOUT BACKBONES, Rev. Ed., Univ. of Chicago Press, 1959. A very clear and readable college text.

Goldring, W. HANDBOOK OF PALEONTOLOGY FOR BEGINNERS AND AMATEURS, Paleontological Res. Inst., Ithaca, N. Y. An older guide with emphasis on New York State.

Moore, Lalicker & Fischer, INVERTEBRATE FOSSILS, McGraw-Hill, N. Y., 1952. A complete and authoritative advanced college text.

Shrock and Twenhofel, PRINCIPLES OF INVERTEBRATE PALEONTOLOGY, McGraw-Hill, N. Y., 1953. A modern, comprehensive reference.

TYPICAL INVERTEBRATE STRUCTURES

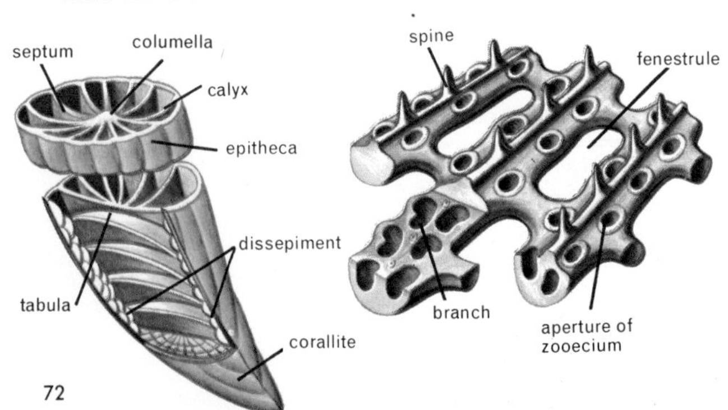

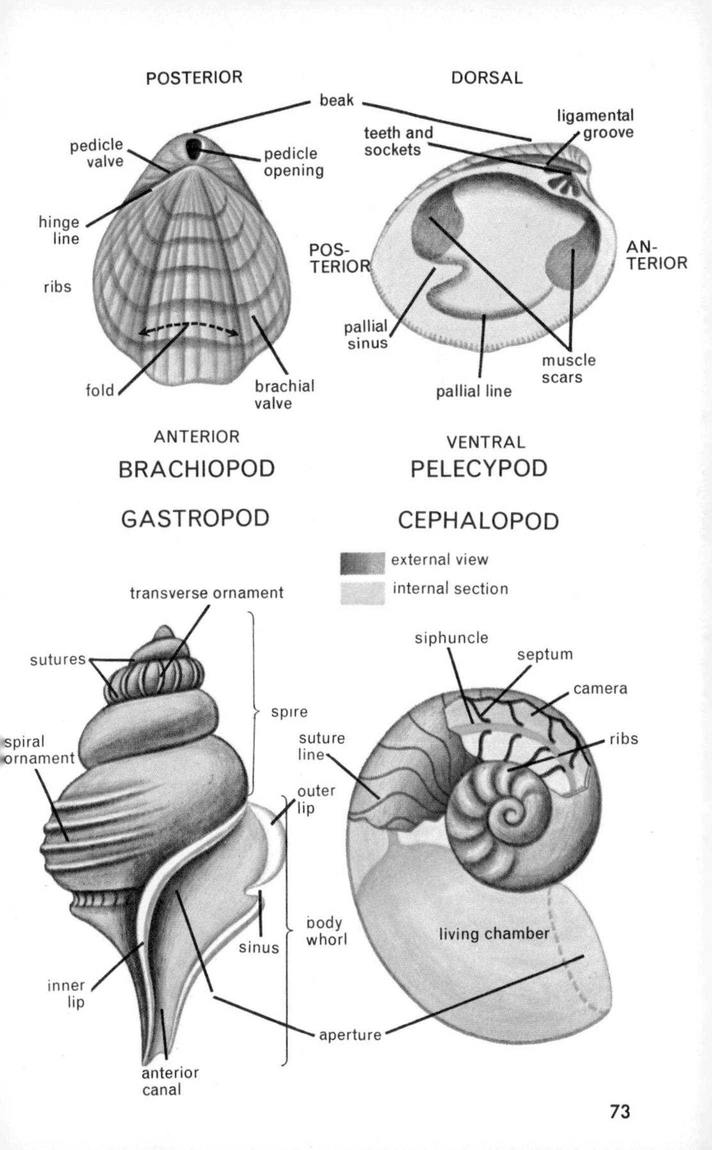
POSTERIOR
beak
pedicle valve
pedicle opening
hinge line
ribs
fold
brachial valve
ANTERIOR
BRACHIOPOD
DORSAL
ligamental groove
teeth and sockets
POS-TERIOR
AN-TERIOR
pallial sinus
muscle scars
pallial line
VENTRAL
PELECYPOD
GASTROPOD
transverse ornament
sutures
spire
spiral ornament
outer lip
body whorl
sinus
inner lip
anterior canal
aperture
CEPHALOPOD
external view
internal section
siphuncle
septum
camera
suture line
ribs
living chamber

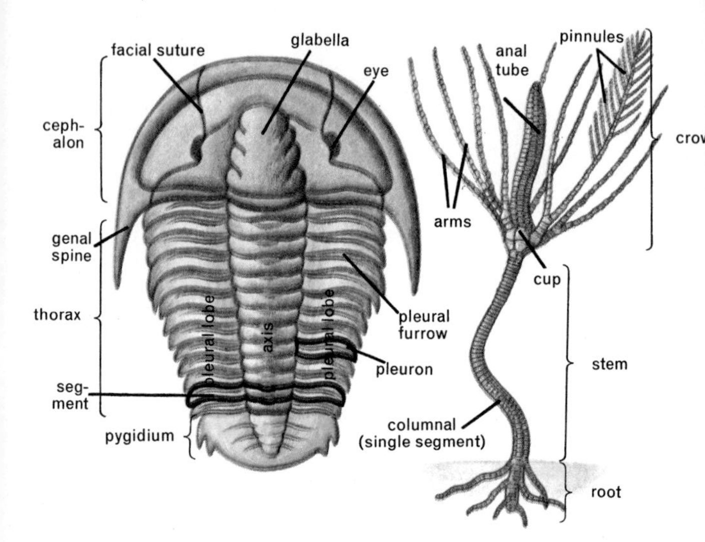

TRILOBITE

CRINOID

ECHINOID

GRAPTOLITE

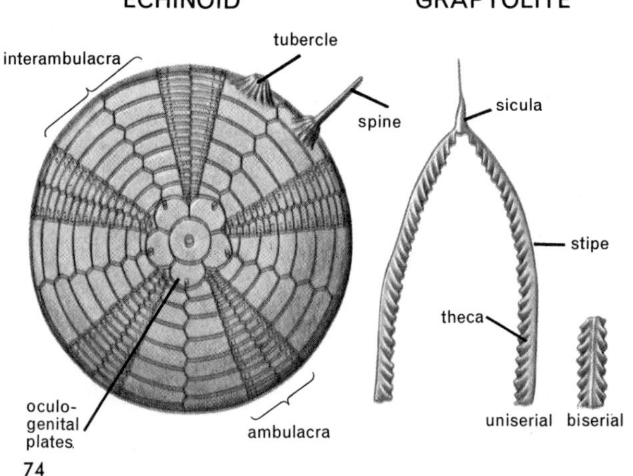

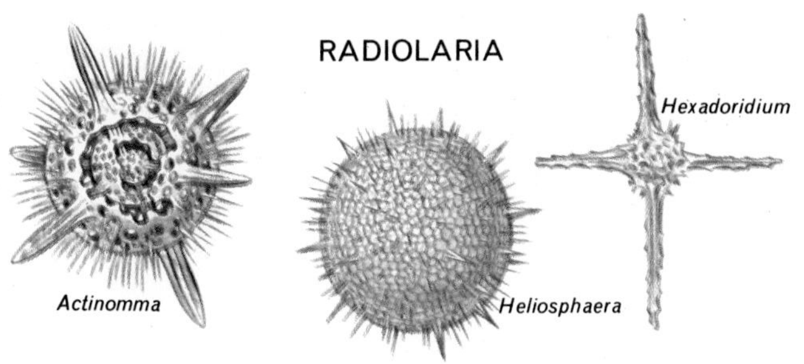

PROTOZOANS are minute, aquatic or parasitic animals whose single cell performs all the life functions. A few species are visible, but most are microscopic. Of the many different groups of protozoans, only foraminifera and radiolaria (Class Sarcodina) are common as fossils. Foraminifera (Ordovician-Recent) usually have tests or shells of cemented foreign particles, or of lime. The shells of radiolaria (Cambrian-Recent) are of silica or of strontium sulphate.

Foraminifera (illustrated below) and radiolaria (above) are so abundant that their tiny shells cover thousands of square miles of ocean floor and form great deposits of ooze. Some limestones are composed largely of foraminifera. Petroleum geologists use foraminifera to identify and correlate strata.

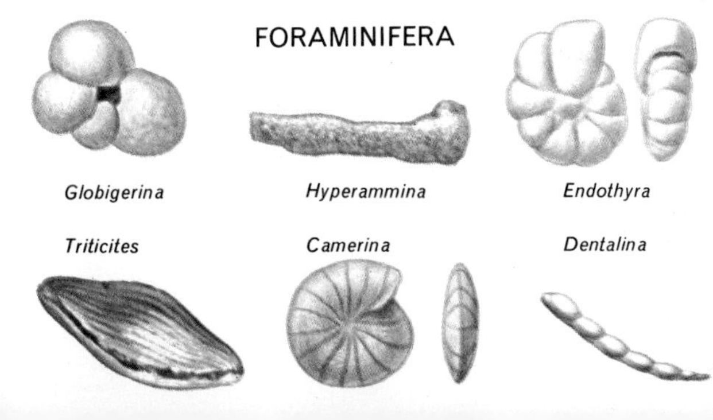

Arrows show water passage through sponge.

cilia
spicules
pore
collar cells
longitudinal section

STRUCTURE OF A SIMPLE SPONGE

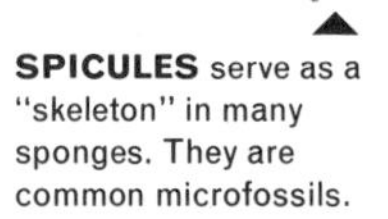

SPICULES serve as a "skeleton" in many sponges. They are common microfossils.

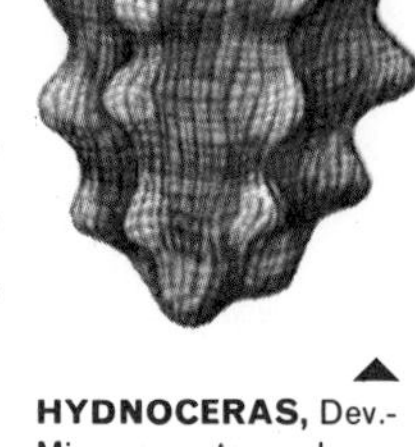

HYDNOCERAS, Dev.-Miss., an octagonal, conical silica sponge. Length about 5 in.

SPONGES or Porifera (pore-bearers) are the simplest multicellular animals. Water is drawn into the sac-like body through many minute pores and is swept along by the whip-like flagella of the collar cells. Microscopic food particles are taken from the water, which is expelled through the top opening of the sponge.

The sponge's support comes from a skeleton of needle-like spicules, made of lime or silica, or from a flexible skeleton of spongin, as in the bath sponge. Spicules are common microfossils. Most sponges are marine, growing attached to the bottom. Some may be as small as a pinhead, others more than 40 inches long. Fossil sponges are found in Cambrian to Recent rocks.

ASTRAEOSPONGIA, Silurian. Saucer-shaped sponge, with prominent spicules. Diameter 2 in.

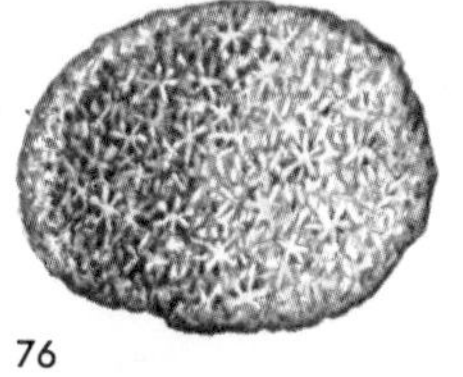

VENTRICULITES, Cret. Conical sponge, with irregular, perforated walls. Length 4 in.

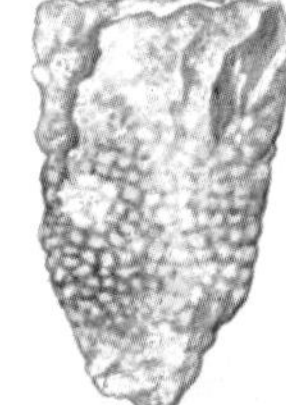

RECEPTACULITES, Ord.-Dev. Globular to dish-shaped. Affinities obscure. Diameter 6 in.

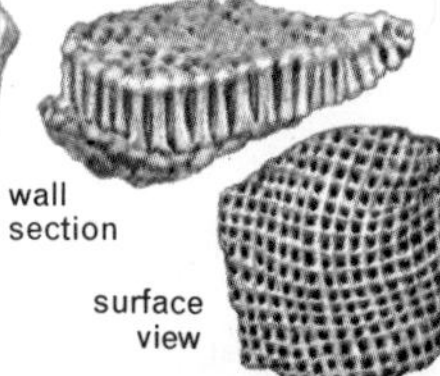

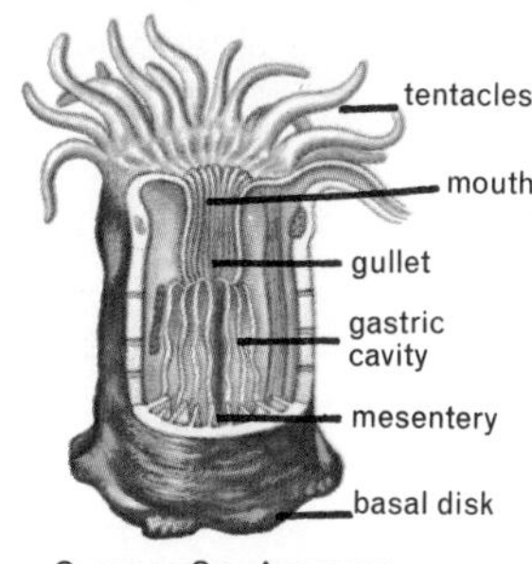

Common Sea Anemone

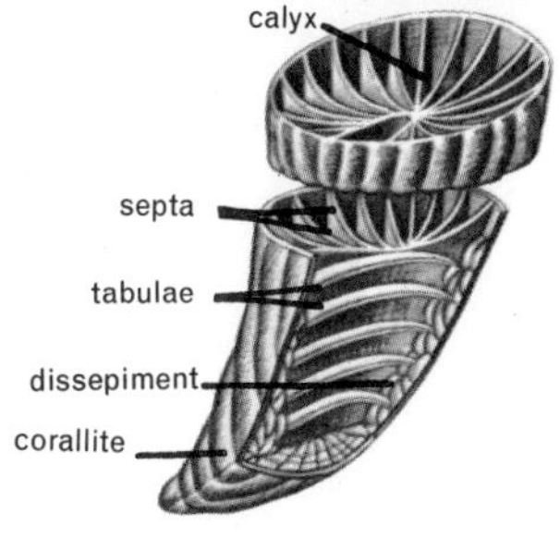

Fossil Coral

COELENTERATES are simple aquatic animals including corals, sea anemones, sea pens and fans, the tiny hydra and the extinct stromatoporoids. Most coelenterates live in the sea; many are colonial. Their sac-like bodies have a two-layered wall and a single opening surrounded by tentacles. They have stinging cells (nematocysts) but lack advanced organ systems, such as the respiratory, excretory, circulatory and nervous systems of higher animals. Jellyfish lack hard parts, but corals and their kin secrete horny or limy "skeletons." The body plan of coelenterates has a radial symmetry. Their reproduction often alternates a sexual stage, the free-swimming medusa, with an asexual stage, the attached polyp.

Corals are important rock builders. Many tropical islands are wholly or partly coral limestone. Living reef-building corals are confined to an equatorial belt of warm, shallow waters. Most require a minimum water temperature of 70°F. and do not grow on muddy bottoms or, because of darkness, at depths greater than about 150 feet. Fossil coral reefs are more widely distributed. Some of them are important petroleum reservoirs. Their presence in both the Arctic and Antarctic indicates climatic conditions very different from the present. Coelenterate fossils are found from the Pre-Cambrian to Recent.

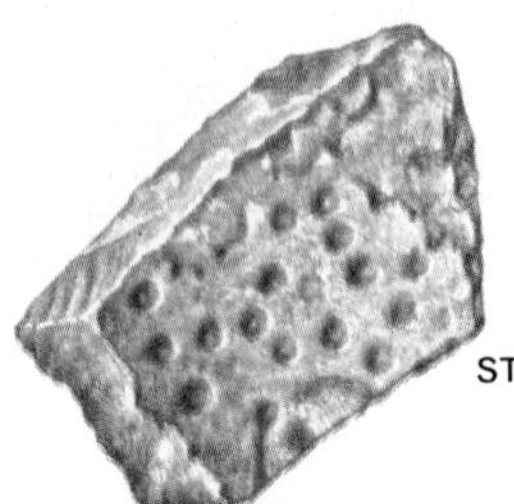

STROMATOPOROID

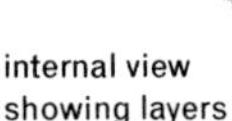

internal view
showing layers

external view showing pores

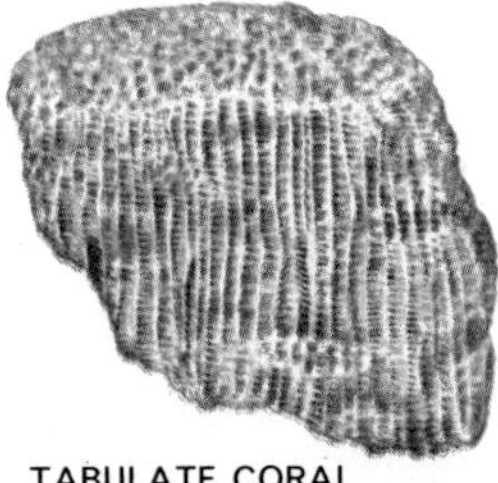

TABULATE CORAL
Favosites

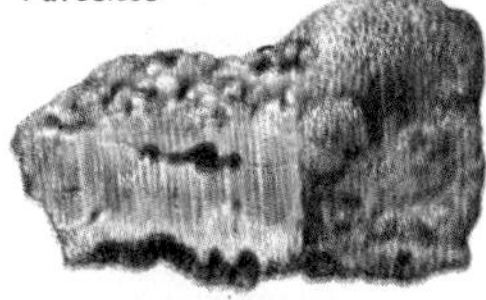

SCHIZOCORAL *Chaetetes*

COELENTERATE FOSSILS include stromatoporoids, extinct sponge-like colonial forms with spherical, branching or encrusting skeletons of lime. Cambrian-Cretaceous.

The corals (Class Anthozoa) include five main groups, three of which are extinct. Tabulate corals (Ord.-Jur.) are compound with strongly developed tabulae, weak or absent septa and no columella. Schizocorals (Ord.-Jur.), often classified under tabulate corals, multiplied by fission and usually lacked true septa. Tetracorals or rugose corals have the main septa developed in four quadrants. Solitary species are called horn corals (Ord.-Perm.). The two living sub-classes of coral are the colonial octocorals (alcyonarians, Trias.-Rec.), with horny or limy skeletons and eight tentacles, and hexacorals (Trias.-Rec.), with a six-fold septal pattern, including reef building corals.

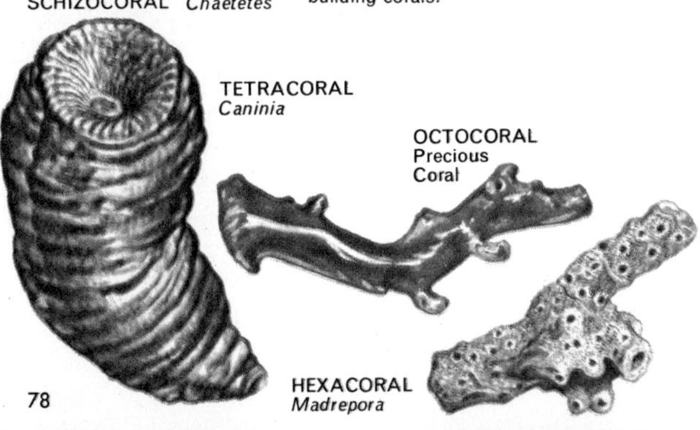

TETRACORAL
Caninia

OCTOCORAL
Precious
Coral

HEXACORAL
Madrepora

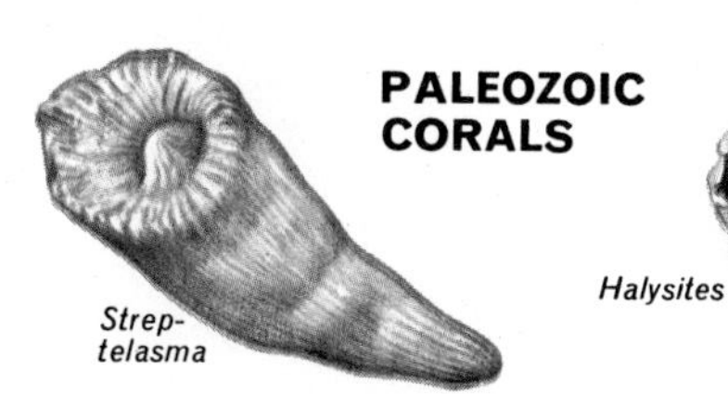
Strep-
telasma

PALEOZOIC CORALS

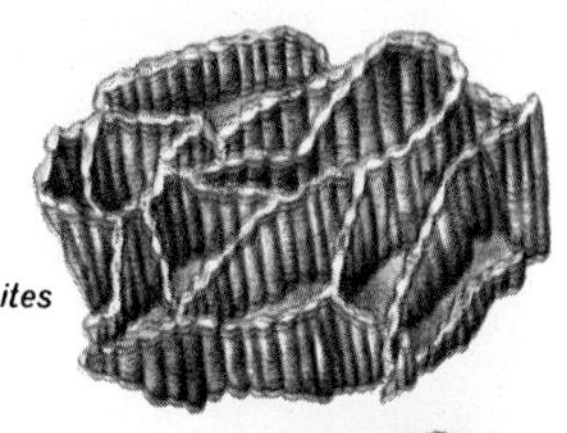
Halysites

STREPTELASMA, Ord.-Dev. Deep calyx. Septa numerous, alternately long and short, thickened at periphery; dissepiments weak. Length about 2 in.

HALYSITES, U. Ord.-L. Dev. Compound "chain-coral," slender corallites arranged in branching rows. Septa weak or absent; tabulae strong. Diameter of colony about 2 or 3 in.

LITHOSTROTION, Miss.-Penn. Colonial coral with cylindrical or prismatic corallites, strong septa, central tabulae, columella and peripheral dissepiments. Max. diam. of corallites about 0.5 in.

CYSTIPHYLLUM, Sil.-Dev. Simple or compound corals of variable shape; interior filled with vesicular dissepiments. Length about 2 in.

SYRINGOPORA, Sil.-Penn., is compound; corallites distinct with transverse connections, tabulae funnel-shaped, septa spine-like. Diameter of colony about 3 in.

LOPHOPHYLLIDIUM, Penn.-Perm. Single coral with projecting columella; septa alternate in length; arched tabulae; no dissepiments. Length about 1 in.

NEOZAPHRENTIS, Miss. Small, solitary coral with one main septum long, opposing one short; oblique fossula on convex side of coral. Incomplete tabulae. No dissepiments. Length about 1 in.

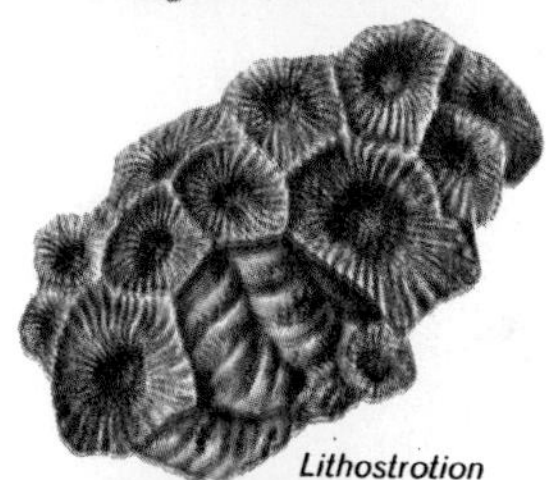
Lithostrotion

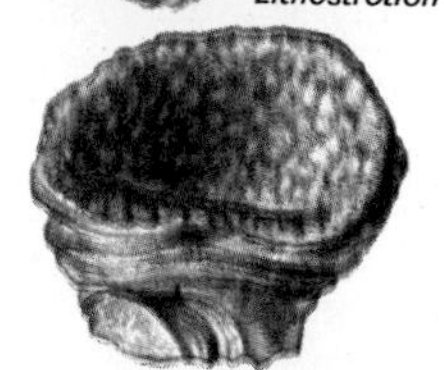
Cystiphyllum

Syringopora

Neozaphrentis

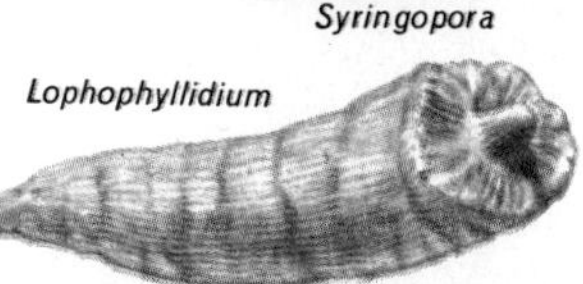
Lophophyllidium

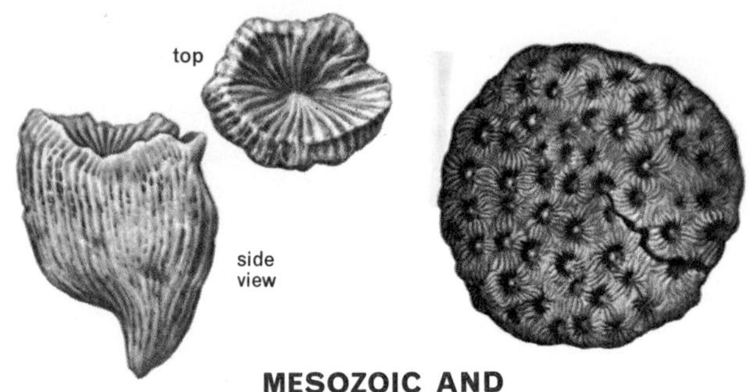

MESOZOIC AND CENOZOIC COELENTERATES

MONTLIVALTIA, Triassic to Tertiary. A single coral, conical or flattened, with a wrinkled surface, numerous septa, outer edges toothed or ridged, numerous dissepiments. Length from 1 to 3.5 in.

STYLASTER. An Eocene to Recent genus of colonial hydrozoan with slender branching and with radiating apertures. Also, pits with bulging margins are present. Length of typical branch 2-3 in.

THAMNASTERIA, Triassic to Tertiary. Compound coral with depressed, flower-like surface; walls of corallites indistinct; septa strong, joining adjacent corallites; small columella. Diam. of colony 3-4 in.

EUSMILIA, Oligocene to Recent. A colonial, branching, stony coral. Septa are prominent and are seen in the walls. No columella. One of an important and numerous group. Length about 1.5 in.

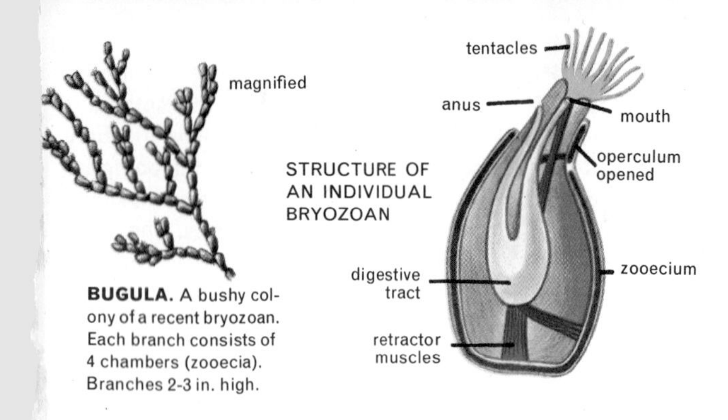

BUGULA. A bushy colony of a recent bryozoan. Each branch consists of 4 chambers (zooecia). Branches 2-3 in. high.

BRYOZOANS or Moss Animals are aquatic, colonial animals with encrusting, branching, or fan-like growth. The supporting structures are horny or limy, and the minute individual animals are housed in tiny cups. Bryozoans resemble corals but are more complex. They have nervous and muscular systems and complete, U-shaped digestive tracts. Most bryozoans are marine and are relatively common fossils. Inconspicuous bryozoans are useful in correlating strata. Ordovician-Recent.

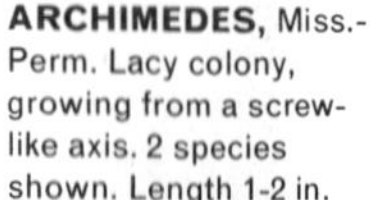
ARCHIMEDES, Miss.-Perm. Lacy colony, growing from a screw-like axis. 2 species shown. Length 1-2 in.

RHOMBOPORA, Ord.-Perm. Slender branches, often spiny; conspicuous apertures. Length about 0.3 in.

FENESTRELLINA, Sil.-Perm. Fan- or funnel-shaped colony; two rows of apertures on one face. Length about 2 in.

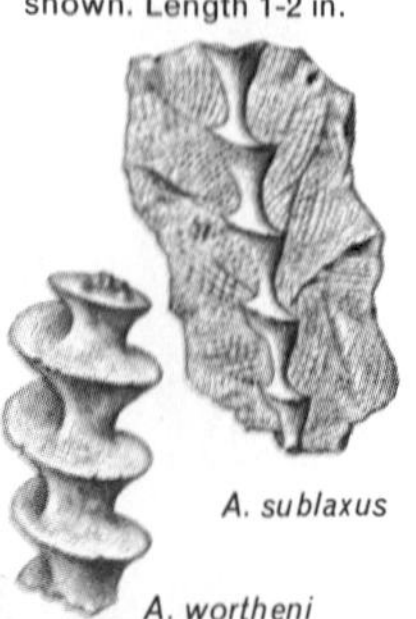

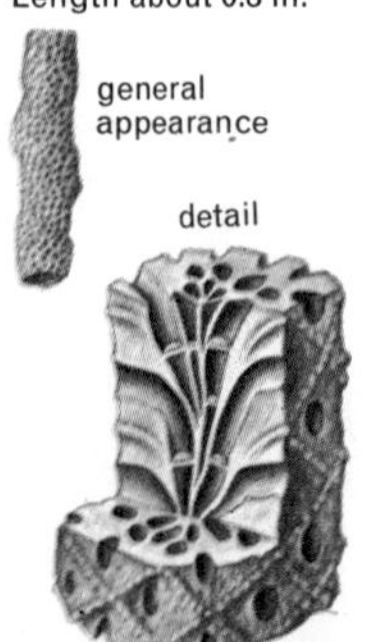

BRACHIOPODS or Lamp Shells are small marine invertebrates. Their variable shells enclose the soft body and other internal structures important in identification. The shell is made of two unequal valves. At the posterior end of one (the pedicle valve) is an opening through which a fleshy anchoring stem (the pedicle) emerges. The other valve is the brachial valve. One group, the inarticulate brachiopods (below), has valves of chitin and calcium phosphate held together only by muscles. The second and larger group, articulate brachiopods, has shells of lime held together by teeth along the hinge as well as by muscles. There are about 200 species of living brachiopods and about 30,000 fossil forms, found from the Cambrian to the present. Cambrian brachiopods are mainly inarticulate. Later, articulate forms became much more common while inarticulate forms declined.

Brachiopods are among the most abundant Paleozoic fossils. Some grew up to 9 inches across. Most were about an inch in diameter. Although adults live mainly attached, they begin life as free-swimming larvae, which

INARTICULATE BRACHIOPODS

LINGULA, Ordovician to Recent. This living fossil, a widely distributed burrowing form, has a thin phosphate shell and a long pedicle. Length about 1 to 1.5 in.

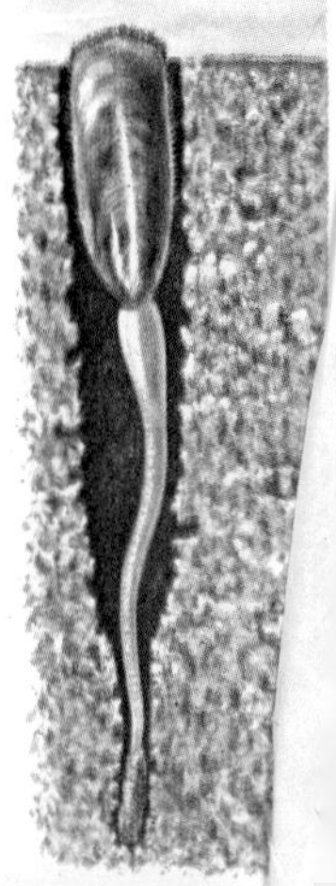

OBOLELLA, Lower Camb.; equal, sub-circular valves marked by fine concentric growth lines. Length about 0.2 in.

LINGULELLA, Camb. to L. Ord. is broadly tear-shaped with a groove inside the pedicle valve. Maximum length 1 in.

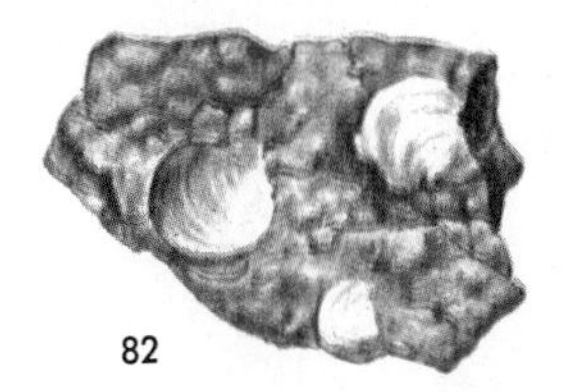

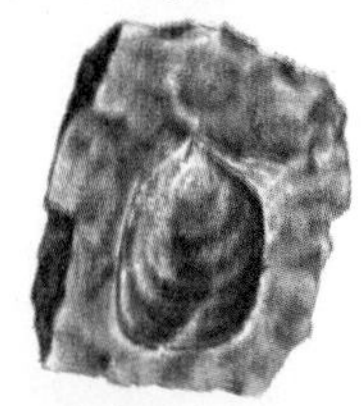

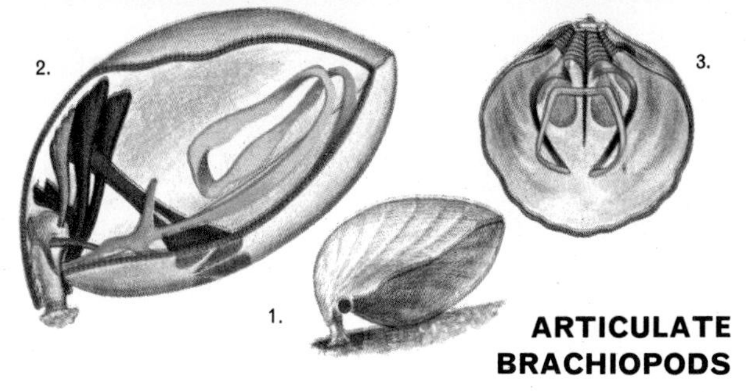

ARTICULATE BRACHIOPODS

Articulate brachiopod, *Magellania,* shows (1) the pedicle valve above and brachial valve below. The red dot marks the end of the hinge line. Interior structures (2) include the brachidia (blue), muscles (red), muscle scars (gray) and pedicle (yellow). The interior of the brachial valve (3) shows the looped brachidia (blue), muscle scars (gray), and hinge sockets (black). Fossils showing interior structure may occur.

accounts for their wide geographic distribution. Lack of a pedicle opening in the shell indicates that some fossil brachiopods did not have a functional pedicle. Nearly all fossil forms lived in shallow water. So do most of the living species, though some have become adapted to greater depths.

PLANES OF SYMMETRY

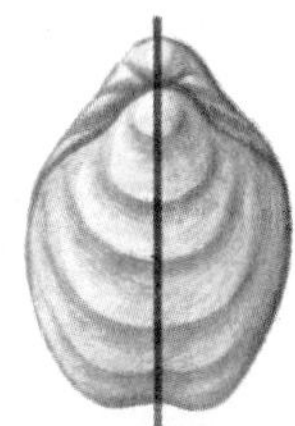

Brachiopod

Pelecypod

Superficially, brachiopods resemble pelecypods (p. 116). Both have 2-sided or bilateral symmetry. In brachiopods a line dividing the shell into two similar halves runs through the valves (as shown above). In pelecypods the line runs between the two valves.

LOWER PALEOZOIC

DINORTHIS, Middle to Upper Ordovician, is an example of a large common group (orthids). All have biconvex profile, a straight hinge line and fine radial ribs. Length 1 in.

pedicle view

profile

pedicle view

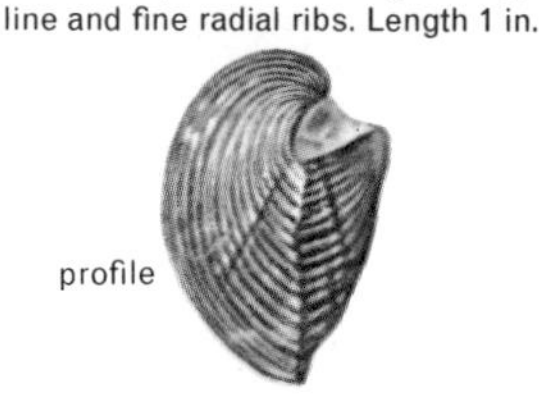

profile

HEBERTELLA, Middle to Upper Ordovician, is a massive shell with pedicle valve rather flat. Brachial valve strongly convex, sometimes with a strong fold; fine radial ribs. Length 1.2 in.

brachial view

profile

pedicle view

ZYGOSPIRA, Middle Ordovician to Lower Silurian, is a small, strongly ribbed shell with a subrounded outline and biconvex profile. Pedicle valve deeper than brachial, with a strong fold. Common throughout the Ohio Valley and in the Appalachians. Length 0.3 to 0.7 in.

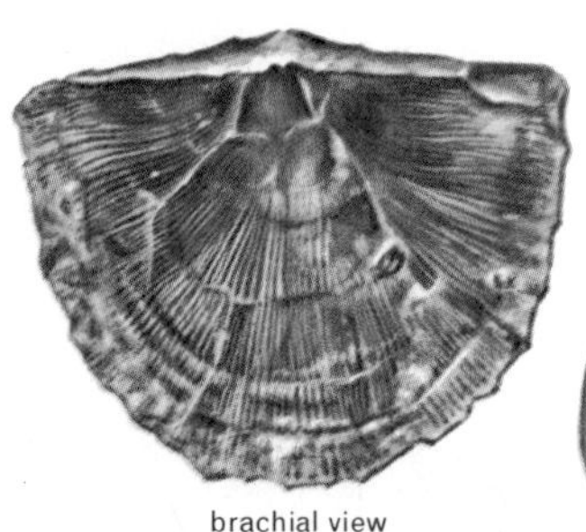

brachial view

profile

RAFINESQUINA, Middle to Upper Ordovician, is a large, flat, semicircular shell with a long, straight hinge line. Brachial valve is flat or concave; the pedicle valve is convex. Fine ribs, sometimes alternating in size. This brachiopod apparently lost its pedicle at maturity. Length 1.3 to 1.7 in.

BRACHIOPODS

PLATYSTROPHIA, Middle Ordovician to Middle Silurian, has a massive, strongly ribbed shell with a convex outline and a long, straight hinge line. Length to 1.7 in.

brachial view

profile

PETROCRANIA on *Rafinesquina,* Middle Ordovician to Permian. A small, limy inarticulate brachiopod. Pedicle valve cemented (often to another shell); brachial valve low, conical. Length 0.3 to 1 in.

LEPIDOCYCLUS (*Rhyncotrema*), Middle to Upper Ordovician, has a ribbed shell often with "herringbone" pattern. Strongly inflated, sub-circular profile. Short hinge line. Brachial valve has a fold. Particularly common in the Ohio Valley. Length 0.3 to 1.3 in.

pedicle view

brachial view

profile

STROPHOMENA, Middle to Upper Ordovician. Outline similar to *Rafinesquina,* but with a concave pedicle valve and a convex brachial valve. It has its strongest curvature near the anterior edge; fine ribs. Like *Rafinesquina,* it probably lost its pedicle at maturity and rested on the bottom. Length 0.6 to 1.4 in.

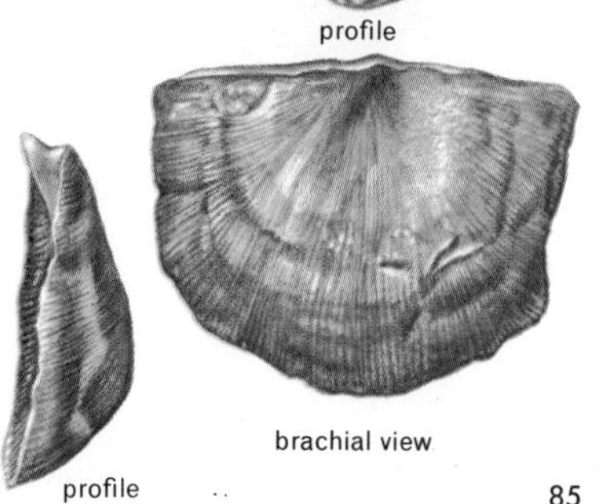
brachial view

profile

LOWER AND MIDDLE

ATRYPA, Middle Silurian to Lower Mississippian. Ventral valve slightly, and brachial strongly convex; ribs variable, sometimes frilled. A widespread fossil. Length 1 to 1.3 in.

profile

brachial view

posterior view

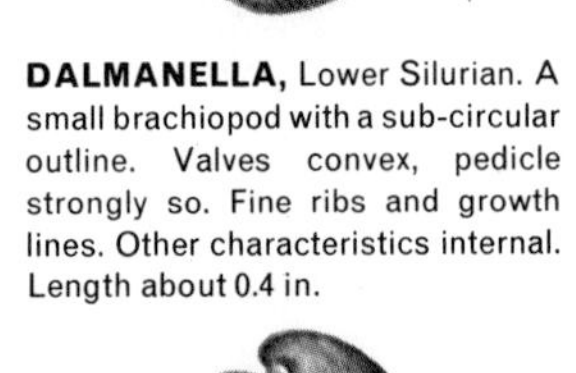

DALMANELLA, Lower Silurian. A small brachiopod with a sub-circular outline. Valves convex, pedicle strongly so. Fine ribs and growth lines. Other characteristics internal. Length about 0.4 in.

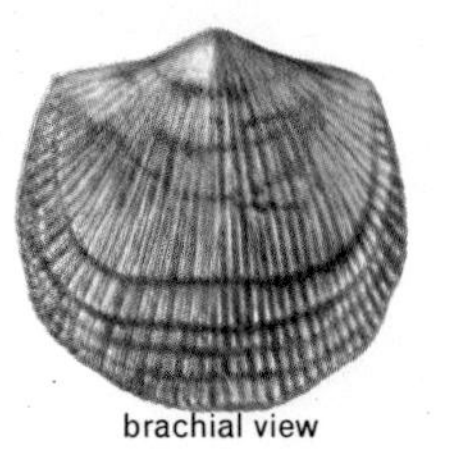

brachial view

profile

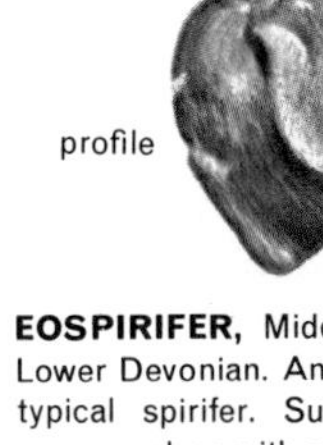

profile

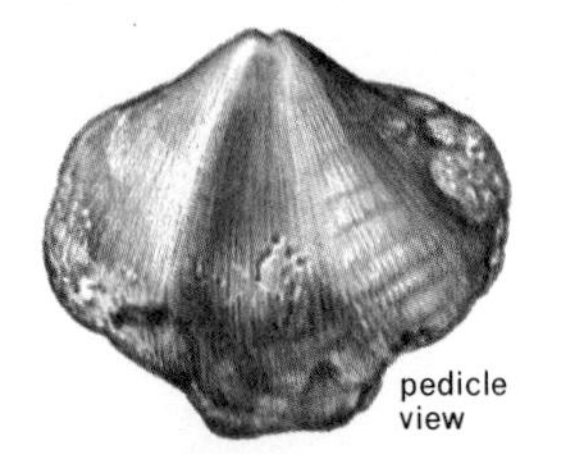

pedicle view

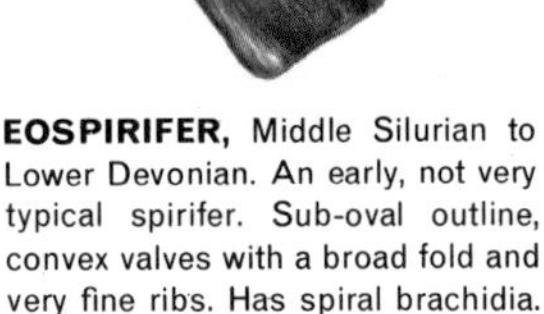

EOSPIRIFER, Middle Silurian to Lower Devonian. An early, not very typical spirifer. Sub-oval outline, convex valves with a broad fold and very fine ribs. Has spiral brachidia. Length 1 to 1.3 in.

DICOELOSIA, formerly called Bilobites, Upper Ordovician to Lower Devonian, has a strongly bilobed outline, narrow hinge and very fine ribs. Length about 0.3 in.

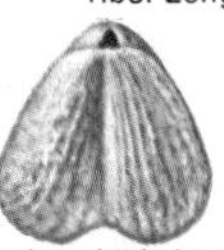

brachial view

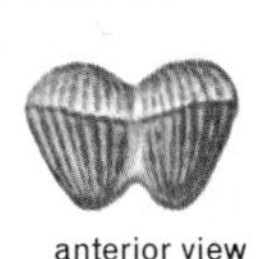

anterior view

two pedicle views

PALEOZOIC BRACHIOPODS

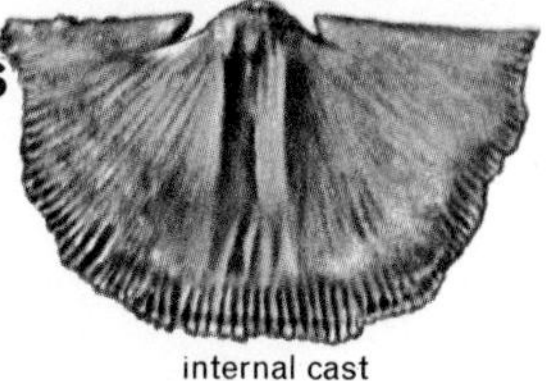

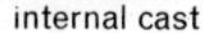

internal cast

SCHUCHERTELLA, Lower Devonian to Permian, is a brachiopod with a wide hinge line; valves flat or gently curved; fine ribs. Other characters are internal. Length about 1 in.

profile

MERISTINA, Middle Silurian, has a large, strongly convex shell with a low fold in the brachial valve. Smooth surface. Other identifying features are internal. Length about 1 in.

brachial view

profile

pedicle view

PENTAMERUS, Middle Silurian. A large, strongly convex shell with smooth valves and a strong beak. Interior casts, as illustrated, show a strong vertical plate in the pedicle valve. Max. length 3 in.

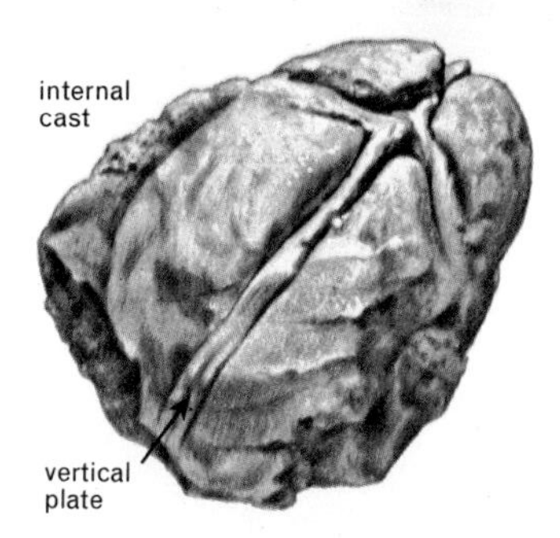

RHYNCHOTRETA, Silurian, is triangular in outline with a sharp beak, prominent pedicle opening and very strong ribs. Length 0.5 in.

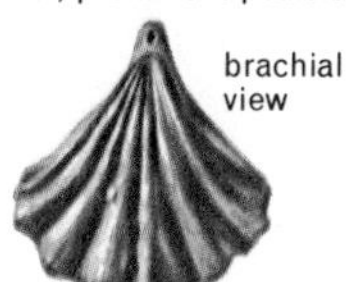

brachial view

profile

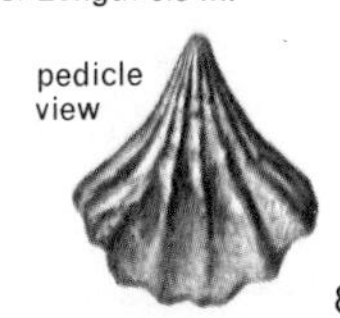

pedicle view

BRACHIOPODS

pedicle view

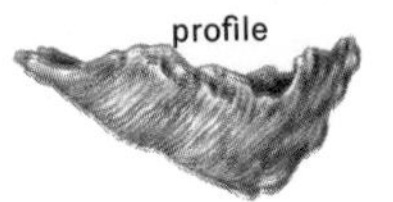

LEPTAENA is a common and widespread brachiopod of Middle Ordovician to Mississippian age. The pedicle valve is convex and the brachial valve is concave, giving the shell an unusual profile. Valves are strongly bent near the anterior edge and along the sides. The surface is covered with very fine radiating ribs and concentric growth lines. Length 1 to 1.5 in.

STROPHEODONTA, Devonian. Semi-circular outline with a wide hinge. Brachial valve is gently concave, pedicle valve convex; fine ribs. Has large muscle scars and other features internal. Maximum length about 1 in.

CYRTINA, Middle Silurian to Lower Mississippian, has a large, flat, triangular area between hinge line and beak of arched pedicle valve. Large pedicle openings. Small, convex brachial valve. Length 0.7 in.

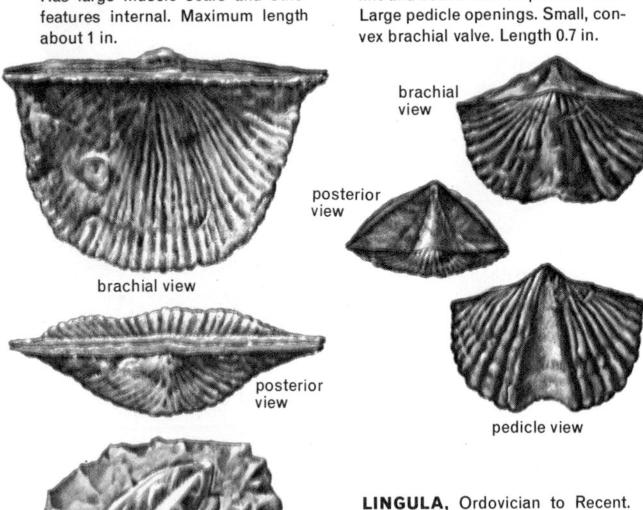

LINGULA, Ordovician to Recent. Inarticulate brachiopod has dark, broadly tear-shaped shell with rather flat profile. Straight anterior margin. See page 82 for living species. Length 1 to 1.5 in.

pedical view

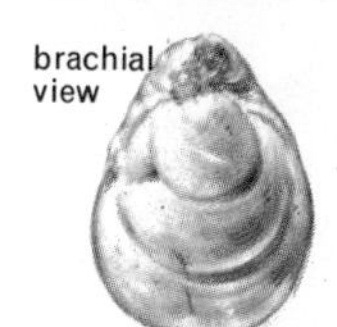

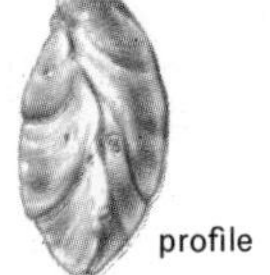

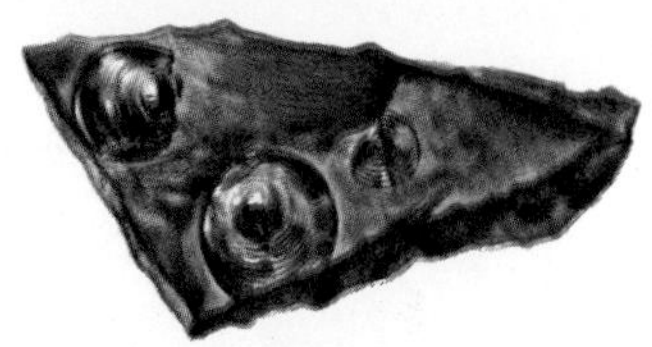

DIELASMA, Mississippian to Permian, has a long oval outline. Both brachial and pedicle valve are gently convex. Surface smooth with few markings. Length to 0.8 in.

ORBICULOIDEA, Ordovician to Permian. Inarticulate. Dark, shiny, conical brachial valve with circular outline and fine rings. The tip of this valve is frequently a bit off-center. Diameter usually about 0.5 in.

SPIRIFER, Mississippian to Lower Pennsylvanian, is the typical member of the spiriferid group of brachiopods, which are common in many Paleozoic strata. Spirifers show great variation in form but have internal spiral brachidia, a more or less triangular outline, and most have radial ribs. *Spirifer* itself has a wide hinge line, strong ribs and a convex profile, with a conspicuous fold. In some species internal structures are important in identification. Length about 1 in.

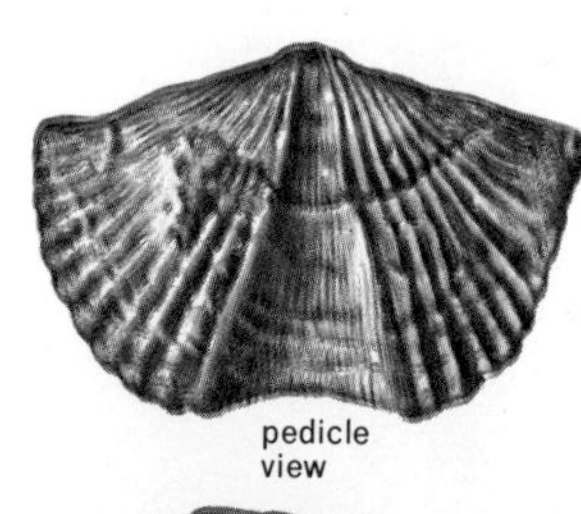

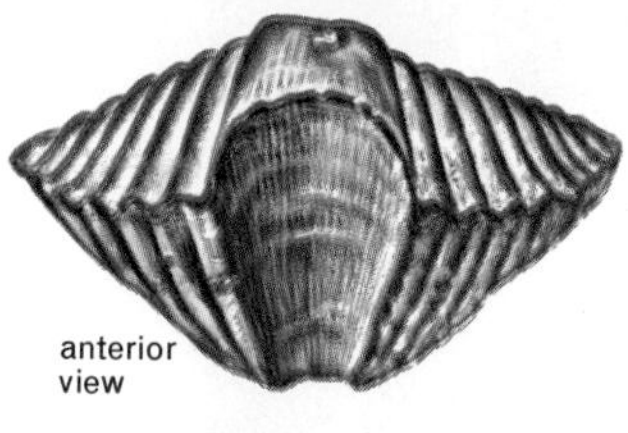

MUCROSPIRIFER, Middle to Upper Devonian, is much wider hinged than *Spirifer,* with shell often winged. Conspicious ribs and fold present; other features internal. Length about 1 in.

brachial view

profile

profile

brachial view

NEOSPIRIFER, Pennsylvanian to Permian, a large, massive spirifer with a wide shell, is convex in profile. Note the prominent hinge, strong ribs, and strong but variable fold. *Neospirifer* is common throughout the mid-continental area. Length 1 to 1.3 in.

pedicle valve

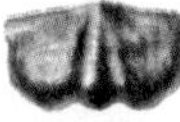
exterior

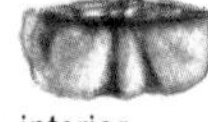
interior

ARTICULATE

MESOLOBUS, Lower Pennsylvanian, has a wide, straight hinge, semi-circular outline, and a depressed profile. Note the shallow fold on the brachial valve and the very fine ribs. Length about 0.3 in.

profile

brachial view

DICTYOCLOSTUS, Mississippian to Permian. Large shells with wide hinge and strong beak. Prominent ribs, with some concentric growth lines and a few spines. Pedicle valve is convex. Length 1 to 1.5 in.

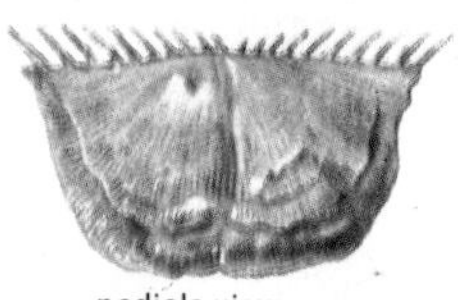
pedicle view

CHONETES, Middle Silurian to Permian, has a semi-circular outline and a wide hinge with spines along margin. Brachial valve concave; pedicle convex. Fine ribs and growth lines. Length 0.4 to 0.9 in.

brachial view

profile

COMPOSITA, Mississippian to Permian, is sub-circular in outline and convex in profile. The shell is fairly smooth with fine growth lines. Pedicle valve has a sulcus. Maximum length about 1 in.

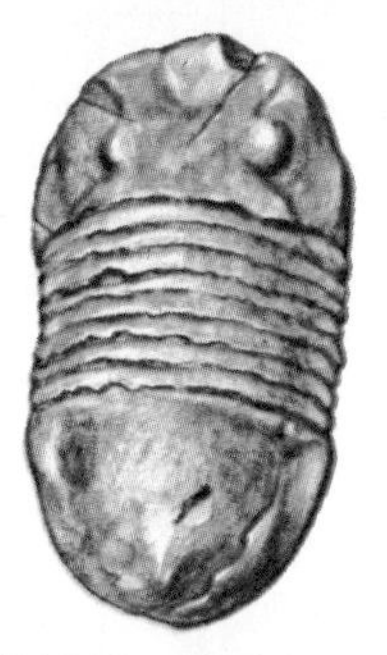

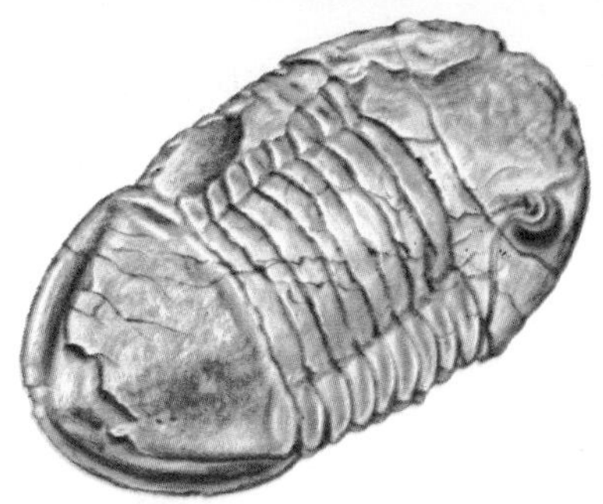

PALEOZOIC TRILOBITES

BUMASTUS, Ordovician to Silurian, has an oval outline; wide axial lobe. Globular cephalon and pygidium lack segmentation. Kidney-shaped eyes. Maximum length about 4 in.

ASAPHUS, Ordovician (an "enrolled" specimen), has a large semicircular cephalon and indistinct glabella; 8 thoracic segments, grooved but blunt. Broad axis. Length about 3 in.

GRIFFITHIDES, Mississippian, has an oval, smooth outline; glabella expanded forward and furrowed. Eyes small; 9 thoracic segments. Pygidium with 13 to 16 segments. Length to 2 in.

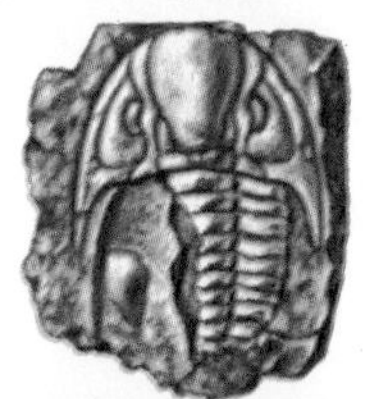

ISOTELUS, Ordovician, has both cephalon and pygidium smooth and sub-triangular in shape; glabella blunt and unfurrowed. Thoracic segments with wide axis. Length about 4 in.; a few forms much larger.

PHACOPS, Silurian to Devonian, has a semi-circular cephalon with rounded corners and inflated glabella. It has 11 grooved thoracic segments. Length usually 2 to 3 in.

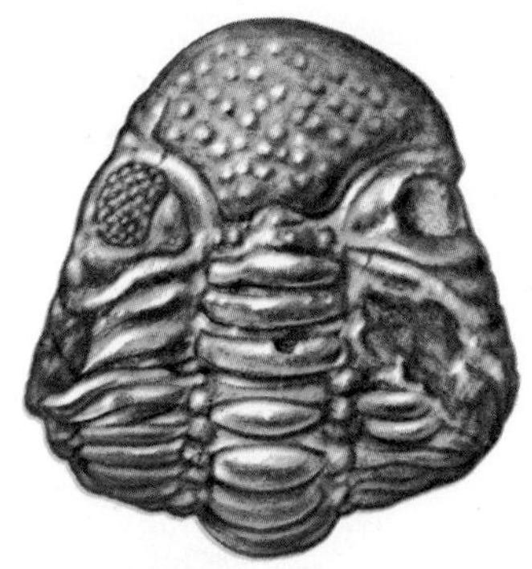

Note the large, raised multi-faceted eyes.

CRUSTACEANS are a major group of marine, fresh-water, and land arthropods. Marine forms, most common and important today, include crabs, lobsters, shrimps and barnacles. Crayfish live in fresh water; sow bugs on land. Crustaceans develop through several larval stages, molting their shells as they grow. All have two pairs of antennae; most breathe through gills. Ostracods are important fossil crustaceans.

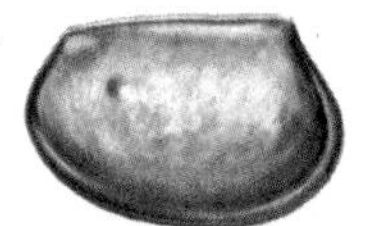

LEPERDITIA, Lower Silurian to Upper Devonian. Large, oblong shell; straight, short hinge; valves unequal. Length about 0.4 in.

DREPANELLA, Middle Ordovician to Lower Silurian. Straight hinge; long marginal ridge and two or more isolated lobes. Length 0.1 in.

OSTRACODS are small, bivalved crustaceans abundant in oceans but also living in fresh water. The animal occupies a laterally compressed shell, hinged along the upper margin. One valve often overlaps the other. Shells have lobes, pits, spines and ribs. Ostracods molt as they grow, and immature molt stages often occur as fossils. Because they are common, ostracods are widely used in correlation of strata of Cambrian to Recent Age.

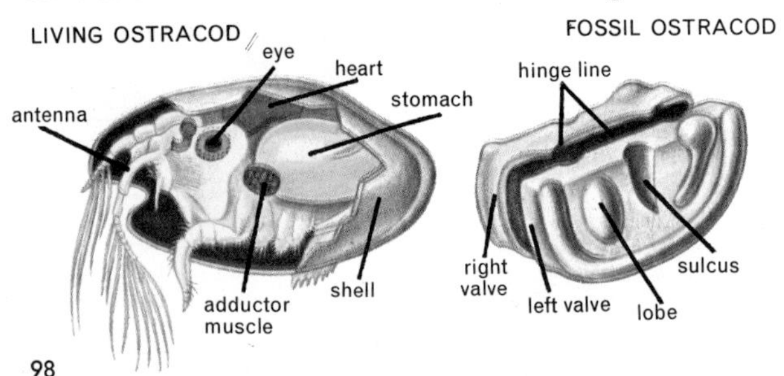

JURESANIA, Pennsylvanian to Permian, has a shell which is semicircular in outline, with a straight hinge. Beak is prominent; pedicle valve strongly convex; brachial valve concave. Spines or spine scars are found on both valves. Maximum length about 1.5 in.

BRACHIOPODS

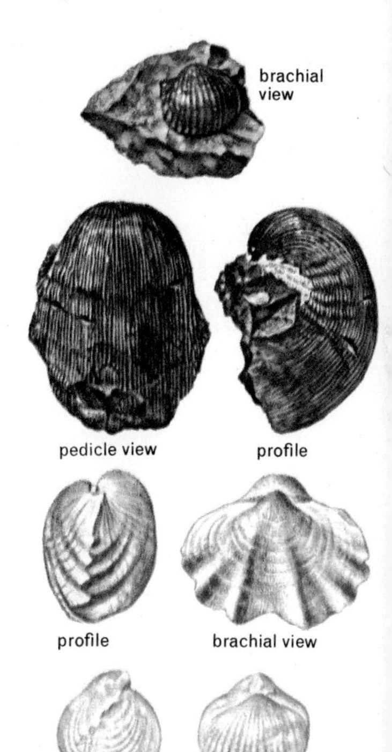

MARGINIFERA, Miss.-Perm., is a small shell with a wide hinge. Pedicle valve strongly convex; brachial valve concave. Medium ribs. Shell may have some spines and a prominent sulcus. Length 0.5 to 1 in.

LINOPRODUCTUS, Miss.-Perm. Very long pedicle valve, often wrinkled near the hinge and strongly curved near the beak; flatter at front. Ribs are prominent and sinuous. Spines may be present. Length about 1 in.

ENTELETES, Pennsylvanian to Permian. A small shell, globular in outline and in profile. The surface of the shell is wavy with very fine ribs superimposed. Other features are internal. Length about 0.5 in.

RHYNCHONELLA is the overall name for a group of common brachiopods (rhynchonellids) of triangular outline and with a short hinge. Most in the group have a convex profile and strong ribs and a sulcus. Length 0.5 to 1.5 in. Ordovician to Recent.

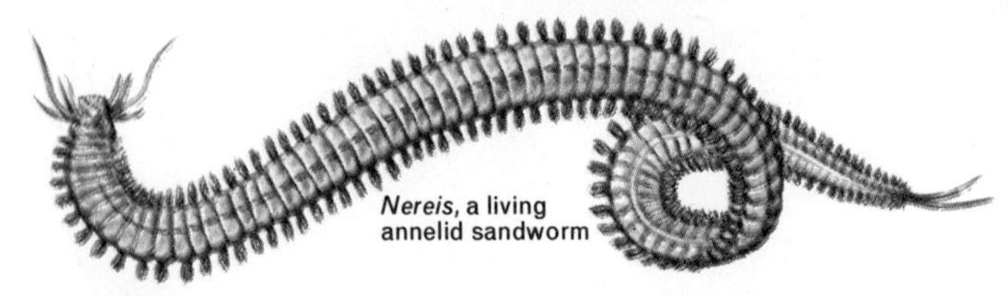

ANNELIDS (above) and related worms are uncommon fossils. They play an important part in geological processes by passing sand and soil through their alimentary tracts. Fossil worm borings, castings, and trails are found. Minute jaw pieces (scolecodonts) of marine worms are common microfossils. Some marine worms secrete limy tubes which may become fossils, and the conical and pyramidal fossils, *Tentaculites* and *Conularia*, may represent remains of worm-like animals. Worms are found in rocks from Pre-Cambrian to Recent.

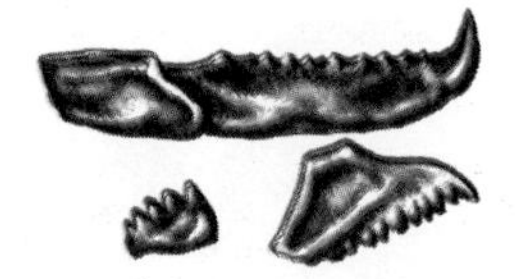

SCOLECODONTS, Ordovician to Recent, are minute chitinous fossils, representing the jawparts of marine annelid worms. Highly magnified.

SERPULA, Silurian to Recent, irregular calcareous tubes, shown here on a brachiopod, are secreted by small worms. Length to 0.2 in.

CONULARIA, length to 6 in., Ordovician to Jurassic and **TENTACULITES,** length to 2 in., Ordovician to Devonian, are obscure extinct forms; possibly worms.

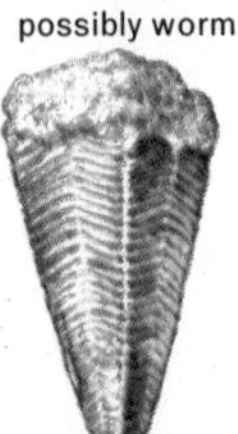

WORM BORINGS are sometimes found in sedimentary rocks. Often they are perpendicular to the bedding of the rocks. Borings are also made by mollusks.

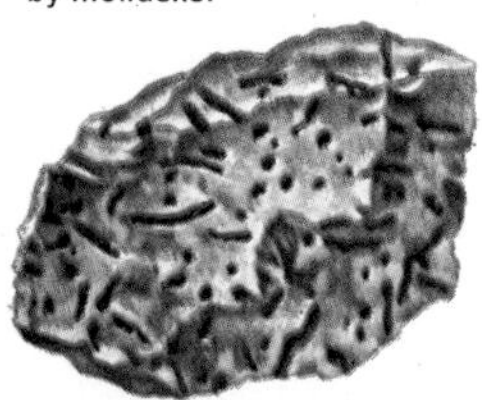

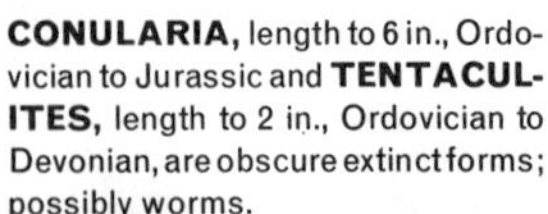

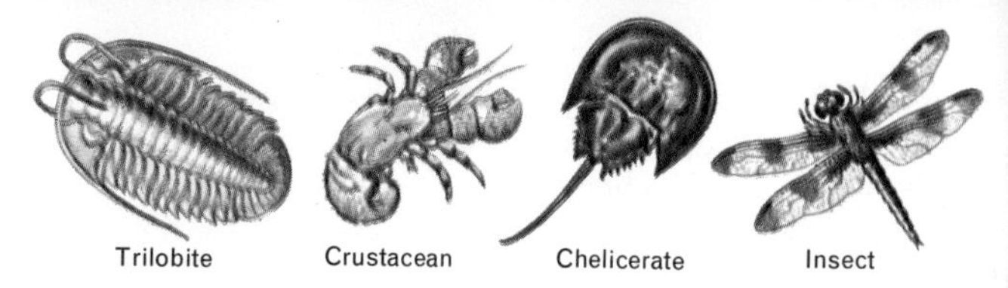

Trilobite Crustacean Chelicerate Insect

ARTHROPODS are a great and successful phylum of invertebrates. They have segmented bodies and paired, jointed limbs, a hard external covering, with flexible joints, and well-developed circulatory, nervous, digestive and reproductive systems. Seven groups of arthropods (spiders, ticks, millipedes and centipedes, lobsters, crabs, barnacles, insects) include over a million species of great diversity. This tremendous phylum is variously divided by experts into classes—the four most important of which are below.

ARTHROPODS IMPORTANT AS FOSSILS

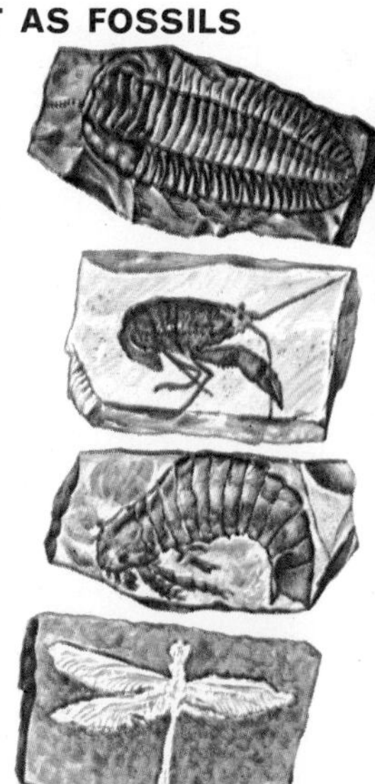

TRILOBITES (pp. 94-97), Cambrian to Permian, are common three-lobed marine arthropods, conspicuously segmented, and have three body divisions. Length ¼ to 27 in. See p. 74 for names of external parts.

CRUSTACEANS (pp. 98-100), Cambrian to Recent. Common, mostly aquatic with two pairs of antennae and usually several pairs of two-branched appendages. Includes crabs, lobsters, shrimps and ostracods.

CHELICERATES (pp. 101-102), Cambrian to Recent. Lack antennae. Appendages may be modified to form pincers. The group includes air breathers (scorpions, mites, ticks and spiders) and water breathers (eurypterids).

INSECTS (p. 103), Devonian to Recent. Winged arthropods, generally with 3 pairs of walking legs. The most-numerous arthropods both in kinds and individuals. Thrive on land and in fresh water.

TRILOBITES, Cambrian to Permian, are extinct marine arthropods of great diversity and importance as Paleozoic guide fossils. The body has three major divisions and the thorax has three lobed segments (see p. 74 for details of structure). They were probably bottom-feeding scavengers and predators.

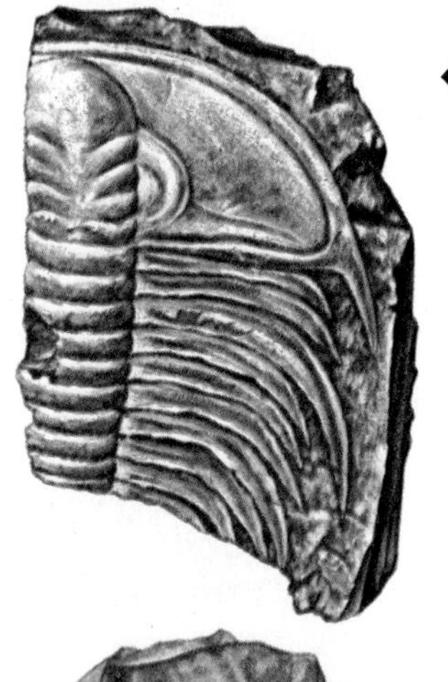

CAMBRIAN TRILOBITES

OLENELLUS, Lower Cambrian. A trilobite with semi-circular cephalon, large crescent-shaped eyes and a long, segmented glabella. The long thorax had many segments; first 15 normal, the rest narrow; some with spines. Length 9 in.

AGNOSTUS, Cambrian. Minute cephalon and pygidium similar in size; no eyes; thorax of two segments. Length about 0.2 in.

CALLAVIA, Lower Cambrian from maritime Canada and W. Europe, has an oval outline; semi-circular cephalon, long and narrow glabella, with long spine. Crescent-shaped eyes. Very small pygidium. Maximum length about 6 in.

ELRATHIA, Middle Cambrian, has an oval outline; semi-circular cephalon with broad, blunt, short glabella; pygidium flat, semi-circular with a smooth margin. Length 1 to 2.5 in.

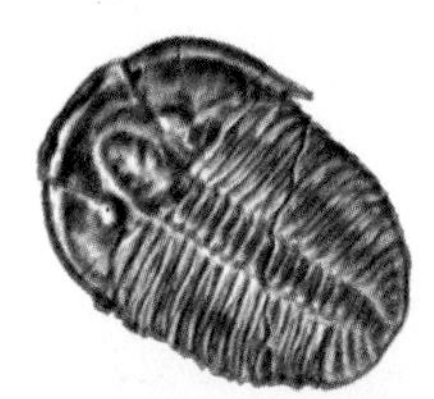

EDRIOASTEROIDS are extinct, attached echinoderms with round or flattened, often asymmetrical bodies covered with small, irregular, flexible plates. Each had a central mouth surrounded by five slender, sinuous, attached arms. Cambrian to Mississippian.

CYSTOIDS, another extinct group of echinoderms, also had rounded bodies made up of many irregular plates with triangular pore openings. They lacked the sinuous arms of the edrioasteroids. Ordovician to Devonian.

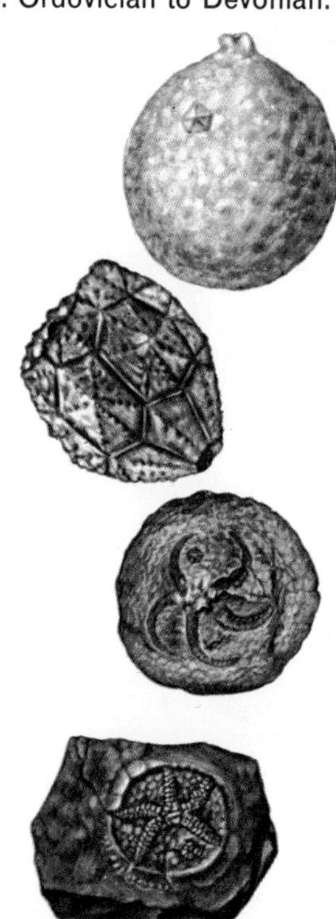

ECHINOSPHAERITES, Ordovician. A globose cystoid of many irregular polygonal plates. Central mouth with short ambulacral grooves. Pyramidal anal covering near mouth. Boundaries between plates often indistinct. Diameter 1 to 2 in.

CARYOCRINITES, Ordovician to Silurian, a cystoid, with a globular body of large, regularly arranged plates. Mouth and ambulacral grooves concealed below plates. 6-13 feeble arms. Pyramidal anal plates; long stem. Diameter about 1 in.

AGELACRINITES, Devonian to Mississippian. An edrioasteroid with five narrow, sinuous ambulacral grooves; three curved right, two left. Periphery has prominent rings of small plates. Diameter about 1.5 in.

HEMICYSTITES, Ordovician to Devonian. An edrioasteroid with a thin, flattened body and five short, radiating arms, surrounded by a ring of large plates and marginal rings of smaller ones. Diameter about 0.8 in.

BLASTOIDS (Ordovician to Permian), an extinct class of echinoderms, are especially common in Mississippian rocks. The fossils consist of a ½- to 1-in., cuplike body which was attached at the base to a short stem. Each "cup" has 13 plates, symmetrically arranged, with five petal-shaped ambulacral grooves. The living animal grew inside the cup as shown on pages 74 and 104.

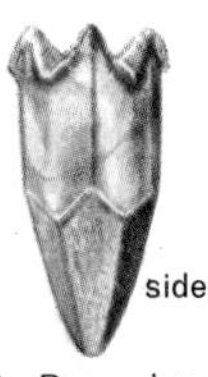

CODASTER, Sil. to Penn., has a pyramidal cup, sharply pointed below and five-sided in cross section. Plates around base and on the sides are long; those near mouth are short. Ambulacral grooves are short and triangular. Height 0.5 to 1 in.

PENTREMITES, Miss. to Penn., has a small, bud-like cup with small basal plates and very long side plates around the broad and petal-like ambulacral grooves. Common in Mississippi Valley. Height about 1 to 2 in.

CRYPTOBLASTUS, from the Mississippian, has globular cup with very long ambulacral grooves which lack pores on outer edges. Plates along the sides are large and overlap those around the mouth. Height about 0.5 in.

SCHIZOBLASTUS, Mississippian to Permian, has an ovoid cup. Plates around mouth are large, with prominent pairs of openings near the top. Base is usually depressed. Ambulacral grooves are long and narrow. Height 0.8 in.

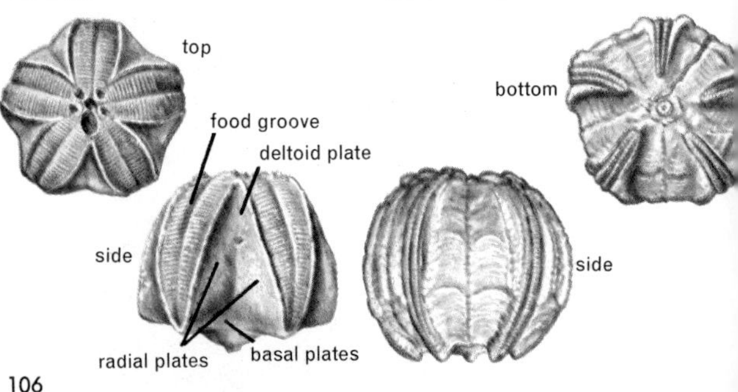

INSECTS, the largest group of arthropods, comprise about three-quarters of all living animals—over 900,000 species are known. Some are enormously abundant and play an important role in human affairs. Insects are adapted to many different environments and are successful in most of them. They have a distinct head, thorax and abdomen, one pair of antennae, three pairs of legs, and one or two pairs of wings. Some are wingless.

Insects are rare as fossils. The oldest are wingless forms from the Devonian. In Pennsylvanian times (Age of Insects) some insects grew to giant proportions and more than 400 species are known. Many living insects show little change from ancient forms in the late Paleozoic. Mesozoic species are more common; over a thousand have been described.

MESOPSYCHOPSIS, a lacewing from the Jurassic. Lacewings (neuropterids), which are distinguished by their fine wing venation, range from Permian to Recent. Length about 1 in.

COCKROACHES (blattoids) are an ancient and widespread group (Pennsylvanian to Recent). Some were up to 4 in. long. 800 Upper Paleozoic species are known.

TARSOPHLEBIA, a Jurassic dragonfly with a wingspan of about 2 in. Outstretched wings are typical of the group, which occurs from Permian to Recent.

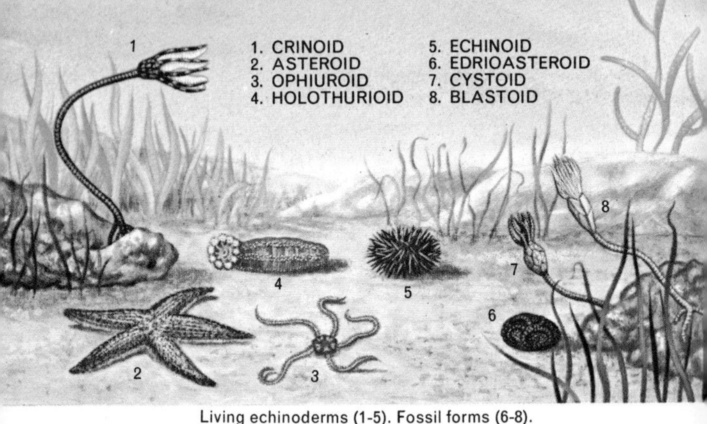

Living echinoderms (1-5). Fossil forms (6-8).

ECHINODERMS are a phylum of marine animals covered with limy plates or spines. Plates are fixed in sea urchins (echinoids), flexible in some starfish (asteroids), and isolated in sea cucumbers (holothurioids). Edrioasteroids, cystoids, blastoids and crinoids lived attached; the rest are free moving. This phylum consists of eight common classes illustrated above. Below are typical internal structures. Digestive system is green; water-vascular system orange. Exoskeleton is purple. These animals also have a well-developed nervous system.

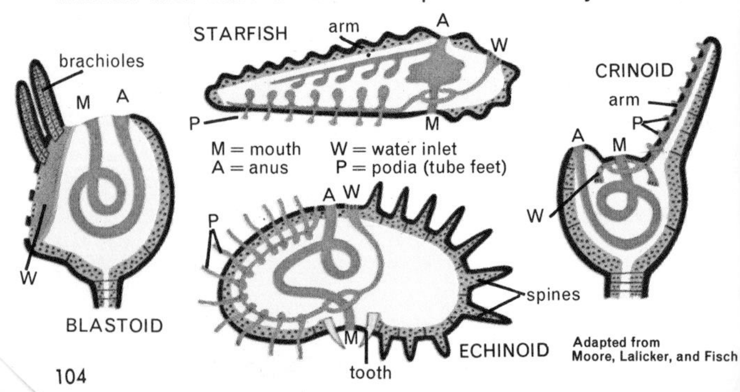

CRINOIDS, or sea lilies, are flower-like echinoderms, often beautifully colored, which grow in colonies on the sea floor. Some fossil forms were free swimming but most were fixed by a stem formed of variously shaped discs or columnals (p. 74) surmounted by a circlet of arms. All have a five-fold radial symmetry, but vary in shape, plates and arms. Ordovician to Recent.

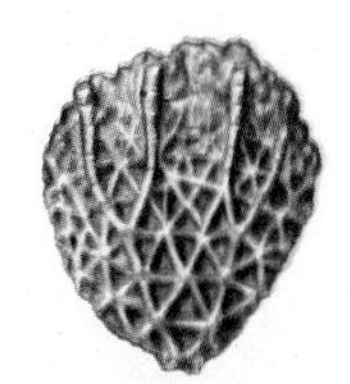

PLATYCRINITES has a deep cup, composed of a few plates, often with a rough surface. The long arms are branched. Mississippian to Permian. Maximum height of crown of this crinoid about 2.8 in., but it is often smaller.

ISOCRINUS, Triassic - Tertiary, and the similar Pentacrinites, have long, branched arms and star-shaped columnals. Both of these stem-bearing, articulate crinoids are characterized by a large crown and a small dorsal cup. Crown to 2.5 in.

GLYPTOCRINUS, Ordovician to Silurian. The cup of this crinoid with its star-shaped ornament is small in comparison with rest of crown. The arms are long, slender and branching. Maximum height of crown 2.5 in.

TAXOCRINUS, Devonian to Mississippian, has a small cup surmounted by massive, branching, embracing arms. The disc-like plates of the stem of this and similar forms are the most common crinoid fossils. Height of crown about 2 in.

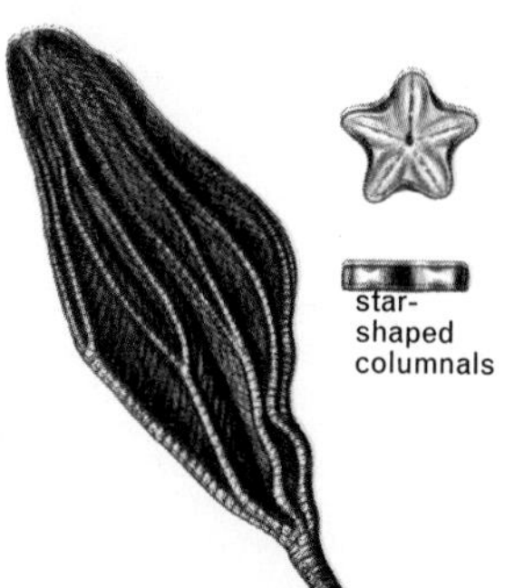

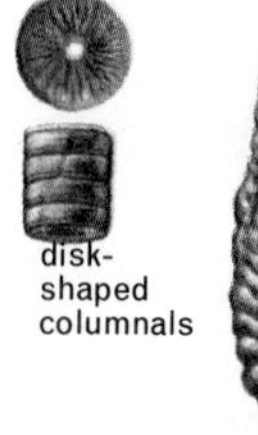

star-shaped columnals

disk-shaped columnals

STARFISH and BRITTLE STARS (Asteroids and Ophiuroids) are free-moving echinoderms. Starfish have five broad arms with tube-feet along the grooves on their lower surface. Brittle stars have a well-defined central disc and slender arms formed of discrete "vertebrae". Both groups are rare as fossils.

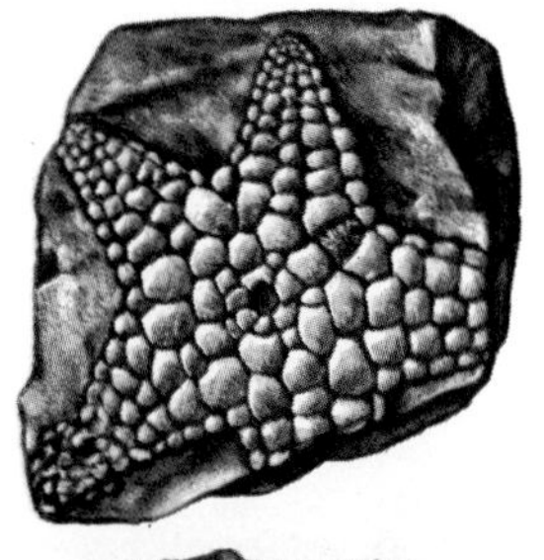

HUDSONASTER, Middle to Upper Ordovician, is an early starfish with thick, short, tapering arms covered with large, regularly arranged plates. It has prominent ambulacral grooves on the lower surface. Diameter about 1 in.

MESOPALAEASTER (DEVONASTER) is an Ordovician-Devonian starfish with sharply pointed arms and regular radiating ornament on the upper surface. It has a central disc with many small plates. Ambulacral grooves prominent. Diameter about 1.5 in.

URASTERELLA is an Ordovician to Pennsylvanian starfish with long, slender, flexible arms but with no obvious central disc. It has prominent ambulacral grooves on the lower surface and small, irregular plates. Diameter about 2 in.

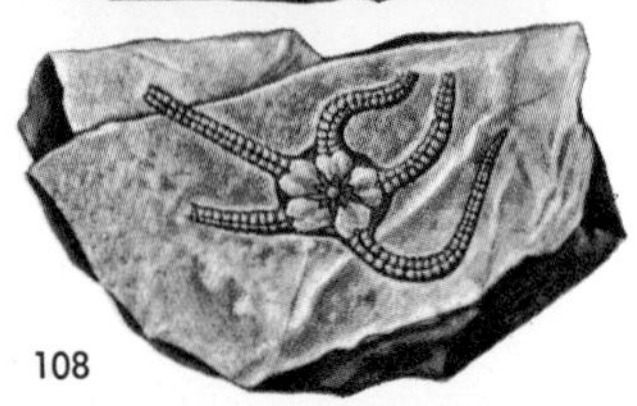

AGANASTER is a Mississippian brittle star with a large, flower-like central disc and short, slender arms. Brittle stars developed in the Ordovician but were not common until the Mesozoic. Often represented by isolated arm plates. Diam. about 1 in.

ECHINOIDS (sea urchins, sand dollars and their kin) have spiny, globular, flattened or heart-shaped shells or tests made of small, limy plates. The test often has five-fold radial symmetry. All Paleozoic echinoids (as *Lovenechinus*) were regular; many later forms (as *Micraster*) are irregular. Found from the Ordovician to Recent.

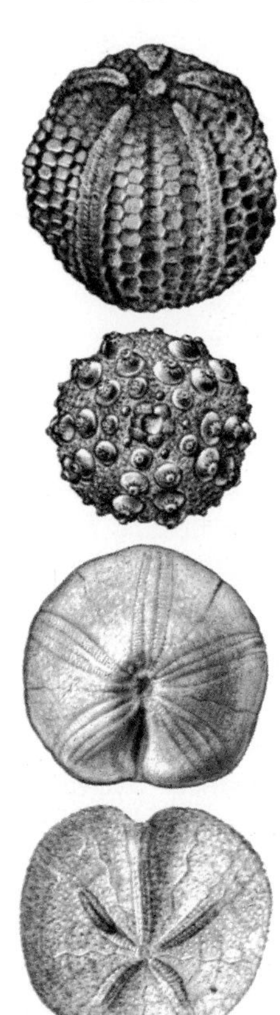

LOVENECHINUS, Mississippian, has a large spherical test, with long, ambulacral areas, each made of four columns of small plates. Between the grooves are four to seven columns of larger plates. Diameter 3 to 4 in.

CIDARIS, Upper Triassic to Recent, is a group name for echinoids with a round test and mouth and anus at opposite poles. Ambulacra long and narrow; the spaces between are broad, with large tubercles and spines. Maximum diameter about 3 in.

CLYPEUS, Jurassic, is an irregular sand dollar with flattened test and a circular or five-sided outline bearing petal-like ambulacral areas. The mouth is at the central point of the test which is covered by very small tubercles. Diam. to 4 in.

MICRASTER, Cretaceous to Miocene, has a thick, heart-shaped test. The ambulacral areas are sunken and the spaces between are filled with large plates. The mouth is near the anterior border. Surface granular. Length about 2 in.

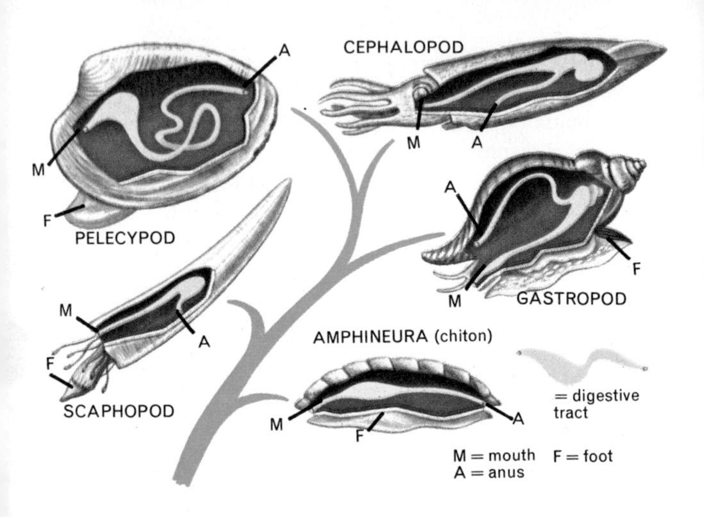

MOLLUSKS include five classes of similar structure but different external appearance, as shown above. Mollusks are an old and successful group; most are marine, many fresh-water, and some live on land. A few aquatic forms are free floating or free swimming but the majority are bottom dwellers in sand or mud. Some burrow into rocks or timbers. About 150,000 living species and thousands of fossil forms have been described. Mollusks range in size from 60-ft. Giant Squids and *Tridacna* clams weighing over 500 pounds down to almost microscopic species. Shell structure varies from the coiled form of snails to the symmetrical bivalves of clams and the eight plates of chitons. Two living classes, the pelecypods (clams, oysters, mussels) and gastropods (whelks, snails, limpets), are abundant. Fossils of these groups and cephalopods are common but chitons and tusk shells are rare, though both occur in the early Paleozoic. Many mollusks are excellent index fossils.

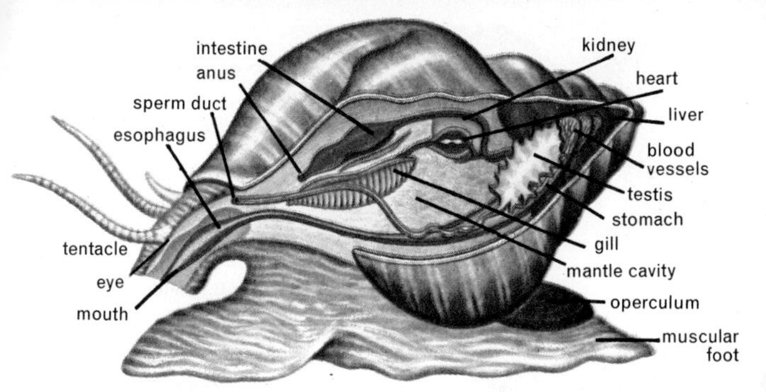

GASTROPODS have a broad, muscular foot, a well-developed head with eyes, mouth, and tentacles. Some have a file-like tongue (radula) which can bore through the shells of other mollusks. Most gastropods have limy spiral shells. Of over 50,000 species, 35,000 are living. The shell opening may be closed by a lid (operculum) when the animal draws in. Once confined to seas, later snails became adapted to life in ponds, in streams and on land.

LOWER PALEOZOIC GASTROPODS

HYPSELOCONUS, Upper Cambrian to Lower Ordovician, is a primitive gastropod with a high, conical, uncoiled shell, the apex of which is often off-center. Form variable. 0.7 in.

HORMOTOMA, Ordovician to Silurian, is a widely distributed, high-spired shell, with whorls rounded and separated by deep notches. Aperture notched; surface quite smooth. 2 to 2.5 in.

MACLURITES is a nearly flat Ordovician gastropod with low but strongly rounded whorls and a broad, deep central depression on the upper side. Surface smooth. 2 to 3 in.

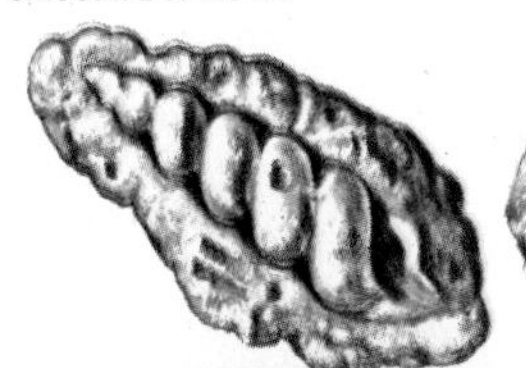

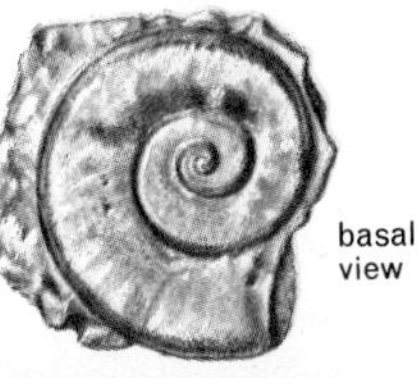

basal view

operculum

PALEOZOIC AND

BELLEROPHON, Ordovician to Triassic. A flatly coiled gastropod with the outer whorl embracing earlier ones. Broad aperture, with sinus. Simple ornament. Maximum diameter about 2 in.

PLATYCERAS, Silurian to Permian. A loosely coiled form often with irregularities in the shell due to cementation to other objects. Spiral or transverse ornament. Maximum diameter about 1.5 in.

PLATYOSTOMA, Silurian to Devonian, is a form of *Platyceras* with a low-spired, globular shell. Several whorls, the last much the largest, are all in contact. Height about 1.5 in.

WORTHENIA, Mississippian to Permian, has a wide, low-spired shell. The outer lip has a shallow sinus which becomes filled in around the whorls. Variable but well developed ornament. Height about 1.3 in.

STRAPAROLLUS or *Euomphalus*, Silurian to Permian, is flatly conical or nearly flat with a broad central depression. Whorls are round to triangular; feeble ornament. Diameter about 2 in.

MESOZOIC GASTROPODS

NERINEA, Jurassic to Cretaceous, is a high-spired, often slender shell. Whorls often concave; aperture with short notch. Variable ornament. Maximum height about 5 in.

MURCHISONIA, Silurian to Permian, is another high-spired shell, rather large, with whorls rounded to angular. Growth lines present. Outer lip with a sinus. Height from 1 to 2 in. (far right).

PLEUROTOMARIA, Jurassic to Cretaceous, has an acutely conical shell with a prominent slit on outer lip and sinus filled in on whorls. Considerable ornament. Maximum height about 2.5 in.

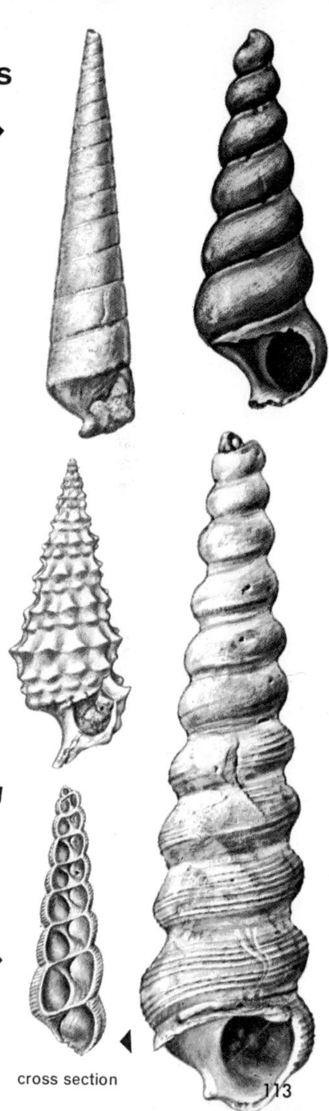

CERITHIUM is a group name for Jurassic to Recent gastropods with high-spired, many-whorled shells, often turreted and with considerable ornament. Maximum height about 5 in.

TURRITELLA, Cretaceous to Recent, is a group name for slender, high-spired shells with incised sutures and with spiral or transverse ornament. Simple aperture. Maximum height about 4 in.

cross section

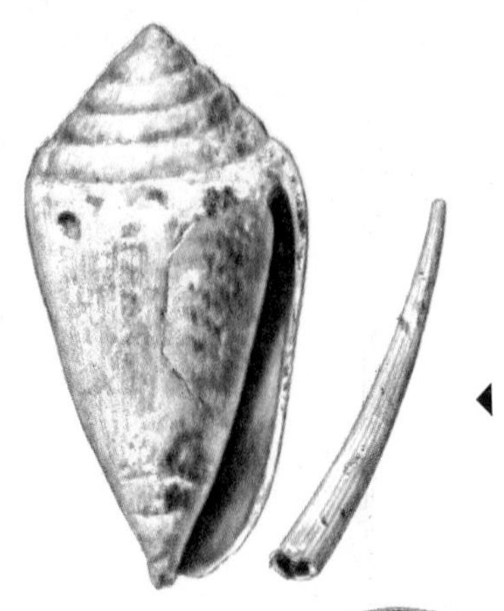

CONUS, Cret.-Rec. A group of acutely conical, short-spired shells. Whorls with straight sides. Long, narrow aperture. Maximum height about 2 in.

DENTALIUM, Ordovician to Recent. A scaphopod (p. 110). Tusk-shaped shell, curved, tapering, with both ends open. Maximum length about 5 in.; usually smaller.

POLYGYRA, Paleocene to Recent. A land snail. Tightly coiled, low-spired shell, flattened below. Indented aperture. Maximum diameter about 1 in.

VOLUTA, Tertiary to Recent. A group of moderately spired shells with angulated, ribbed whorls and narrow aperture. Maximum height about 4 in.

FUSUS, Cret.-Rec. A group of narrow, long shells, with high spire and rounded whorls. Long, narrow anterior canal. Ornament variable. Maximum height about 2.5 in.

VIVIPARUS, Jurassic to Recent. A fresh-water gastropod with a low-spired shell. Whorls are round or flattened and the sutures are indented. Maximum height about 1.5 in.

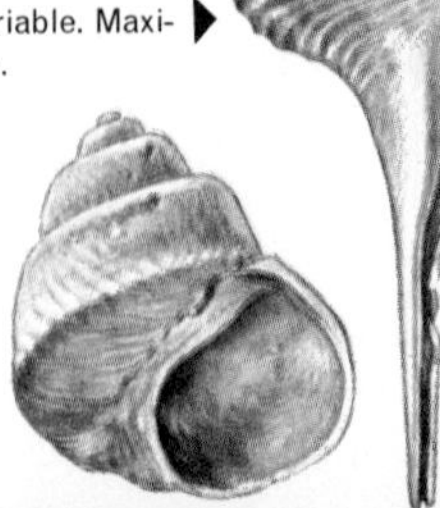

GASTROPODS

PLANORBIS, Jurassic to Recent. A group of fresh-water species. Flatly coiled, with entire spire enclosed in body whorl. Maximum diameter about 0.9 in.

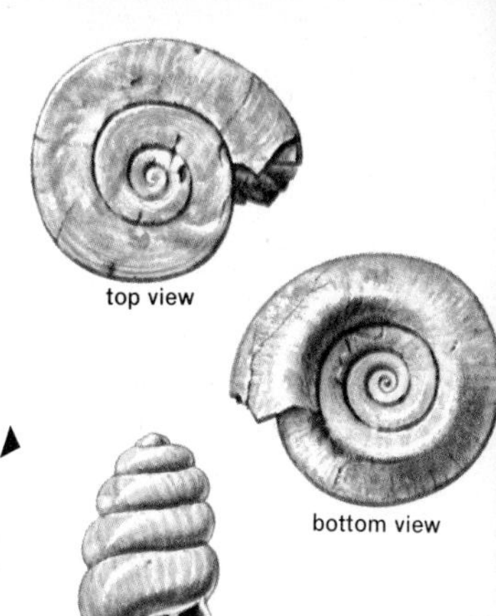

VERTIGO, Eocene to Recent. Land snail with bulbous, oval outline; few smooth whorls; aperture constricted. Height about 0.05 in.

NATICA, Triassic to Recent. A group of low-spired forms, with large, bulbous final whorl; surface generally smooth. Maximum height about 2 in.

CREPIDULA, Upper Cretaceous to Recent. Slipper shells; beak twisted; aperture large, partly covered by a thin platform. Maximum length about 2 in.

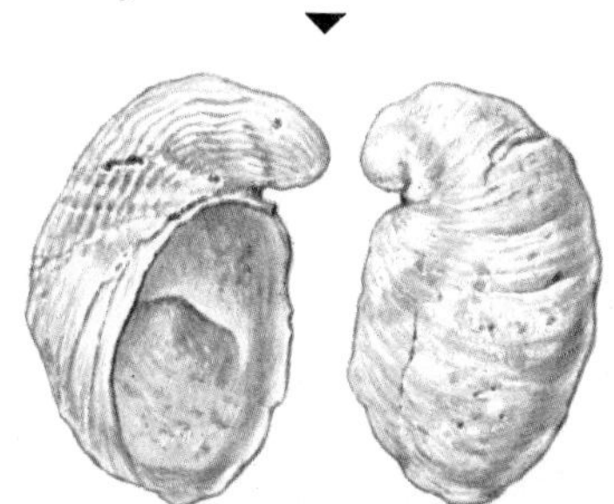

LITTORINA, Paleocene to Recent, is a thick, low-spired and ovoid shell with last whorl larger than rest of shell. Smooth or weak spiral ornament. Maximum height about 1 in.

PELECYPODS, "hatchet-foot" or bivalved mollusks, are mostly marine, but some live in fresh water. Oysters, mussels. and clams are living types. The two valves of the shell, usually similar, are joined along a hinge line and are held together by teeth and muscles, which leave impressions on the inside of the shell. The compressed body of the animal is enclosed by the shell. The shell can open to expose the foot and the siphons by which the animal takes food and oxygen from the water.

Most pelecypods are bottom dwellers but some are active swimmers. Others burrow in sand and mud, some are borers and some cement themselves to fixed objects. Pelecypods (Ordovician-Recent) are common fossils in marine and some non-marine rocks. In Europe pelecypods are known as lamellibranchs, or plate-gilled mollusks. In other areas, all pelecypods are known as clams, though some types may be called mussels, scallops or oysters.

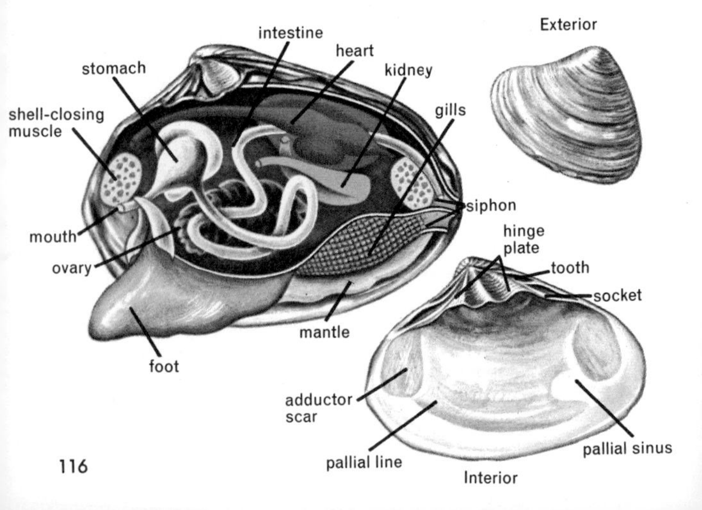

PALEOZOIC PELECYPODS

CTENODONTA has equivalved, oval shells with smooth surfaces, sometimes with fine, concentric growth lines. Numerous similar teeth along hinge plate. Ord.-Sil. Maximum length about 1 in.

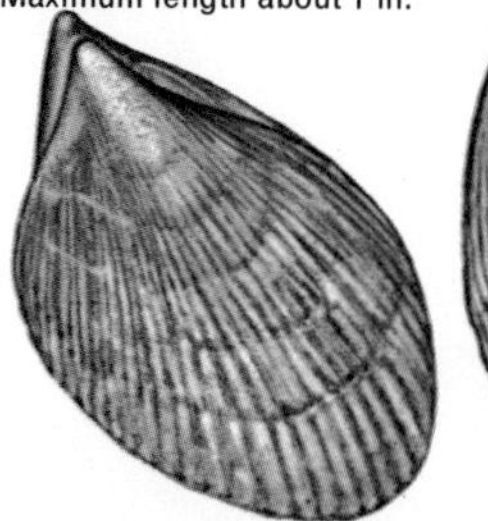

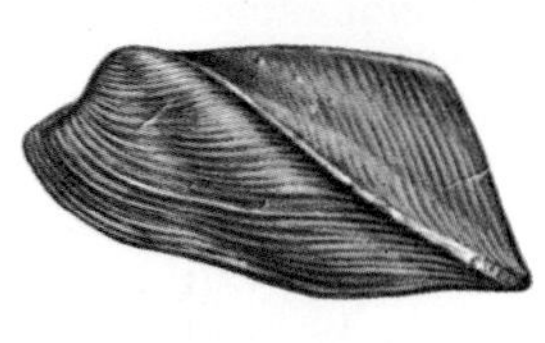

BYSSONYCHIA has a sharp, steeply inclined beak near end of hinge; usually strong radial ribs. These equivalved shells are common in Upper Ordovician. Length about 1 in.

GONIOPHORA, Sil.-Dev. A lopsided shell with prominent beak from which a ridge extends to the rear margin. Maximum length about 2 in.

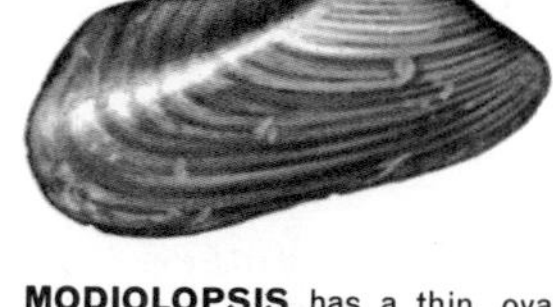

GRAMMYSIA has a prominent, blunted, incurved beak and a suboval outline. A deep, oblique fold runs across the valves. Silurian to Mississippian. Maximum length about 2 in.

MODIOLOPSIS has a thin, oval shell. Asymmetrical valves are crossed by an oblique depression. Ord.-Sil. Two unequal muscle scars. Maximum length about 1.5 in.

PTERINEA has a wide hinge line which extends into "ears." Opposite valves dissimilar and lack symmetry; ornament of fine, concentric lines. It has two unequal muscle scars. Ord.-Penn. Maximum length about 1.5 in.

CONOCARDIUM is a striking fossil of doubtful affinities. Beak prominent; hinge line long and straight. Unequal triangular valves have strong radial ribs and often concentric growth lines. Anterior short with posterior oblique. Ordovician to Permian. Length 1 to 2 in.

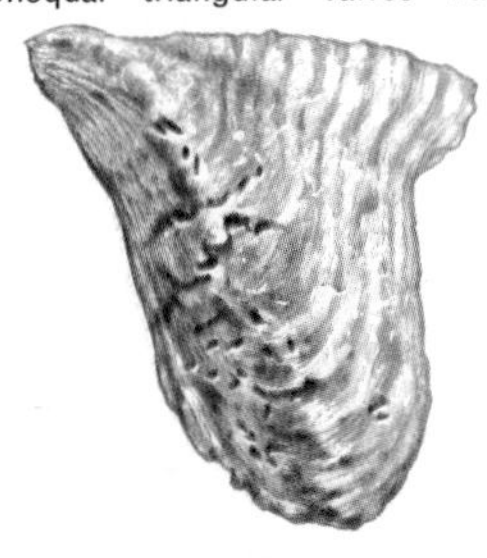

MYALINA, Devonian to Permian, has a pointed, strongly inclined beak, and faint concentric ornamentation. Hinge line often prominent. Length to 4 in.; usually smaller.

ALLORISMA, Miss.-Perm., has an elongated, oval outline. Margin is flattened behind the blunted, anterior beak. Valves gape posteriorly. Maximum length about 2.5 in.

DUNBARELLA are flattened, scallop-shaped shells, with fairly prominent wings and branching ribs. Feeble concentric ornament. Pennsylvanian. Length 1 to 2 in.

CARBONICOLA has inequilateral shaped, elongated valves, oval in outline. A fresh-water genus with high, thick anterior. Ornament of concentric growth lines. Penn. Length 1 to 1.5 in.

PALEOZOIC AND MESOZOIC PELECYPODS

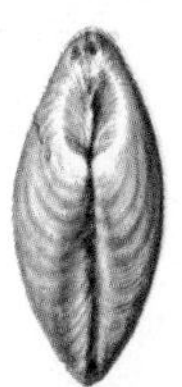

NUCULA is a "living fossil," showing almost no change since Silurian times. Surface commonly has concentric growth lines. Numerous small teeth and sockets. Maximum length about 1.5 in.

AVICULOPECTEN, Silurian to Permian, has a straight hinge with prominent wings and no teeth. Valves unequal, with strong ribs and sometimes growth lines. Length about 1 in.

LIMA is obliquely oval in outline; equivalved and inflated, with radial ribs. The prominent beaks are pointed. Valves often gape. Pennsylvanian to Recent. Length 3.5 in.

TRIGONIA, Jur.-Rec., has a triangular or crescentic outline. Thick valves have a sharp ridge from the beak to the margin. Conspicuous and variable ornament. Maximum length about 3.5 in.

PARALLELODON, Devonian to Tertiary, is angular in outline, elongate, with a long, straight hinge line. Has concentric growth lines. Length about 1.2 in.

PTERIA, Jur.-Rec., has thin, inequilateral shaped valves with a long, straight hinge merging into large, unequal wings. Fine concentric or radial ornament. Maximum length about 3 in.

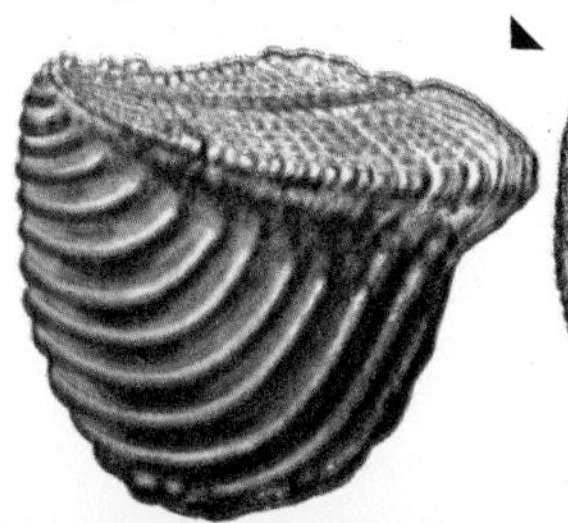

MESOZOIC AND CENOZOIC PELECYPODS

EXOGYRA, Jurassic to Cretaceous, like *Gryphaea* below, but with a large, massive, spirally twisted left valve. One valve was attached and the other served as a lid. Ornament is variable but usually very strongly developed either as growth lines in species to the left above or as ribs in species to the right. Maximum length about 5 in.

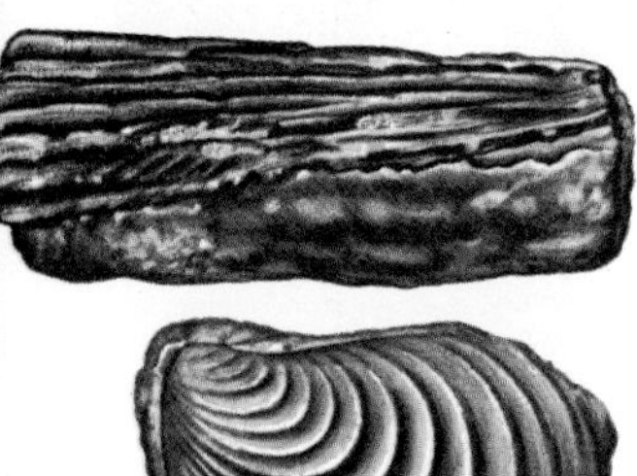

PINNA, Jurassic to Recent. These Pen-shells are triangular, large, thin and equivalved, with a long hinge. The valves are gaping; hinge teeth absent. Attachment to the bottom is by horny threads. Fossils often fragmentary. Maximum length about 9 in.

INOCERAMUS, Jurassic to Cretaceous, is oval in outline, with a prominent beak and a straight hinge line without teeth. It has concentric, corrugated growth lines. Several species have produced fossil pearls. Length up to 4 ft.

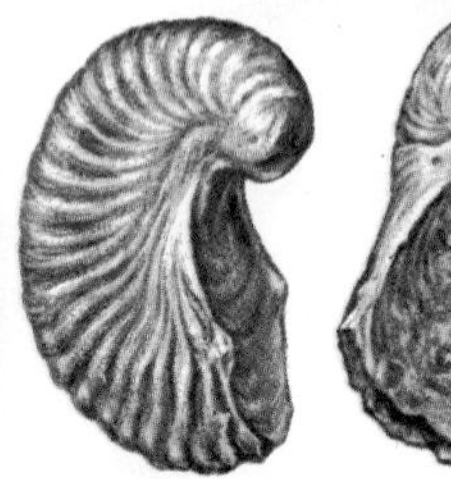

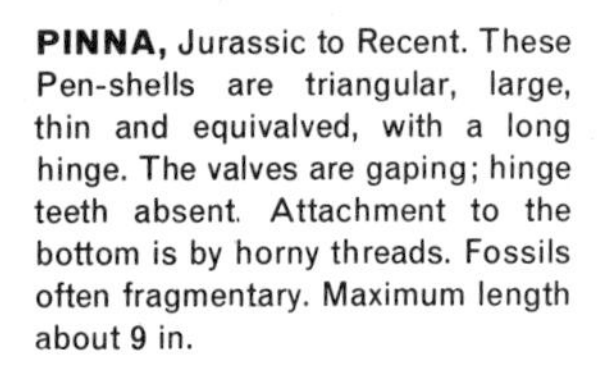

GRYPHAEA, Jurassic to Eocene, is called Devil's toenails. Valves are grossly unequal; left valve loosely coiled, right small, flat and lid-like; growth lines conspicuous. Degree of coiling is variable. Maximum length about 3.5 in.

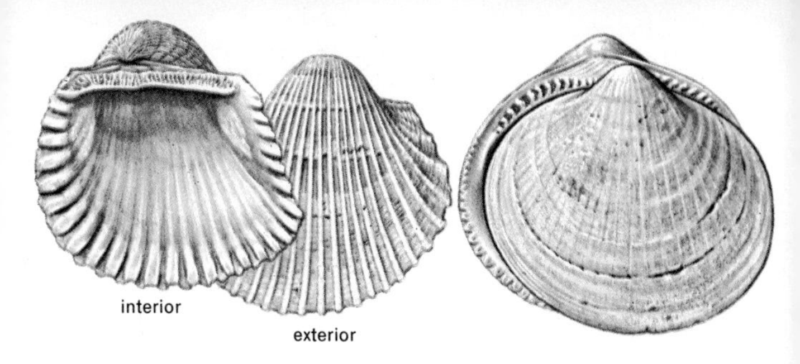

ARCA, Jurassic to Recent, has an angular outline. Beak blunt but conspicuous; teeth and sockets small. Prominent radial ribs. Length 2 to 3 in.

GLYCIMERIS, Cretaceous to Recent. Valves individually symmetrical, nearly circular in outline; pointed beak, with a striated area between beak and hinge. Length 1 to 2 in.

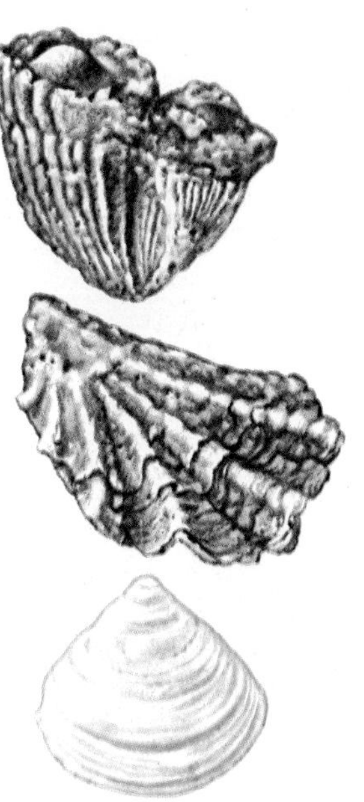

HIPPURITES, Cretaceous, is a widespread conical, coral-like shell. The right valve is very thick, deeply conical and grows attached to rocks. The left valve is thick and lid-like. Valves move on thick teeth. Height about 5 in.

OSTREA, Triassic to Recent. Oysters grow attached by left valve, which is concave, ribbed and larger than right valve, which is flat and often smooth. Ornament may include deep folds and growth lines; shape very variable. Length 2 to 6 in.

ASTARTE, Triassic to Recent, has almost equal valves, is oval to triangular in outline and has a prominent beak. It is smooth or has sculptured growth lines which form concentric ridges. Length about 0.8 in.

CENOZOIC PELECYPODS

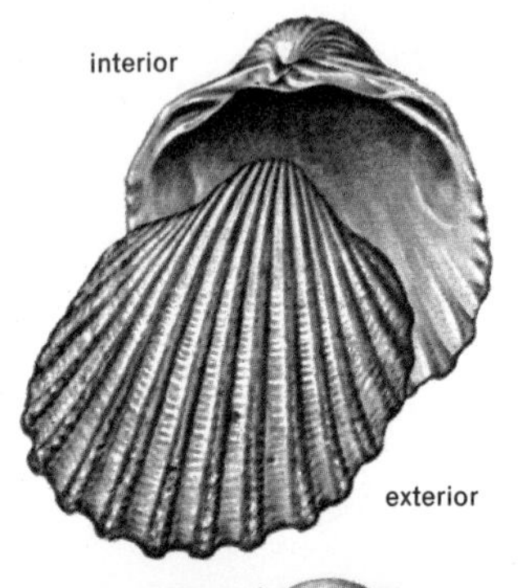

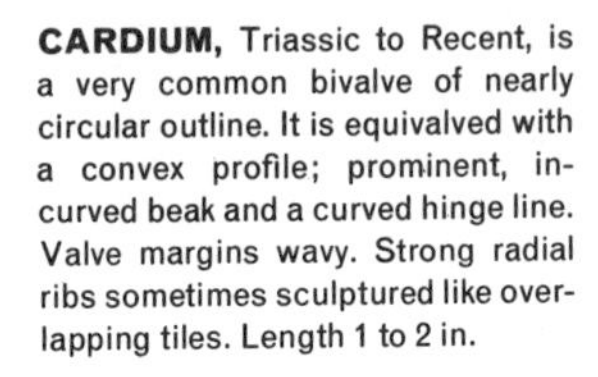

CARDIUM, Triassic to Recent, is a very common bivalve of nearly circular outline. It is equivalved with a convex profile; prominent, incurved beak and a curved hinge line. Valve margins wavy. Strong radial ribs sometimes sculptured like overlapping tiles. Length 1 to 2 in.

UNIO, Triassic to Recent, is a fresh-water clam, oval in outline and equivalved with a blunt but prominent beak. Surface is smooth or has concentric growth lines. Hinge with relatively few, diverse, large teeth. Length 2 to 3 in.

PECTEN, Miss.-Recent, is the group name for many well-known bivalves. Valves individually symmetrical except for unequal wings at end of the long, straight hinge, which has a triangular ligament pit on the inside. Strong radial ribs; single muscle scar. Length 1 to 8 in.

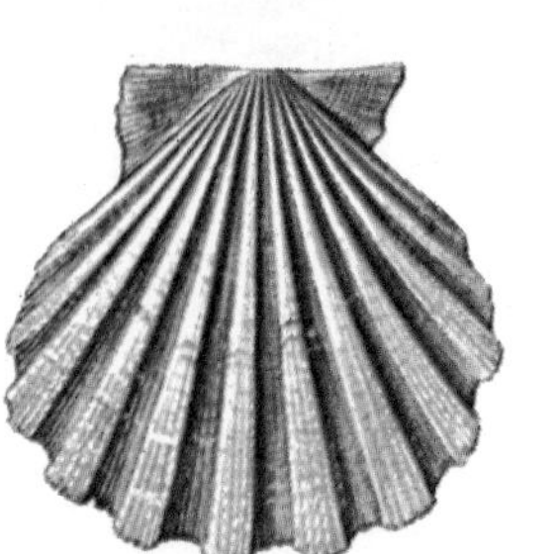

ENSIS, Tertiary to Recent, is the common, widespread razor shell, with an old-fashioned razor outline. The margins are almost straight and the beak is terminal. Ornament is simple, of fine, concentric lines. Length 1 to 10 in.

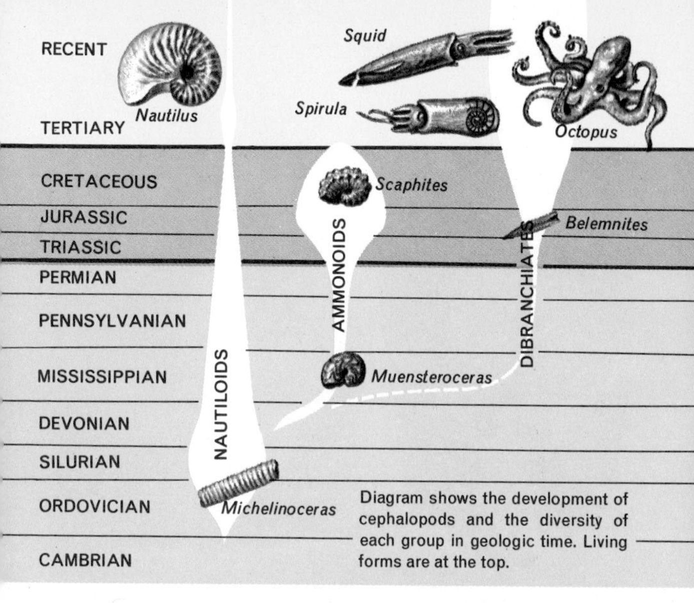

Diagram shows the development of cephalopods and the diversity of each group in geologic time. Living forms are at the top.

CEPHALOPODS are highly developed marine mollusks, represented by the living nautilus, octopus and squid. The shell may be external, internal or absent, and it may be variously coiled. Living forms have a well-developed head, eyes, and tentacles. Most fossil forms had well-developed shells. Three main groups exist. Ammonoids and nautiloids are four-gilled cephalopods with an external shell divided into chambers by transverse plates or septa. The animal lives in the outermost chamber. A fleshy stalk perforates the septa. The junction of the septa with the shell wall forms the suture line.

Coleoid cephalopods (octopus and squid) have two gills and have either an internal shell or none at all. The most common fossil forms, Mesozoic cigar-like belemnites, are the internal skeletons of squid-like species.

EARLY PALEOZOIC CEPHALOPODS

Cephalopods have been traced back to the mid-Cambrian, and in the early Paleozoic had already become widespread. The largest of these early forms reached a length of 15 feet, with straight or gently curved, long shells show-

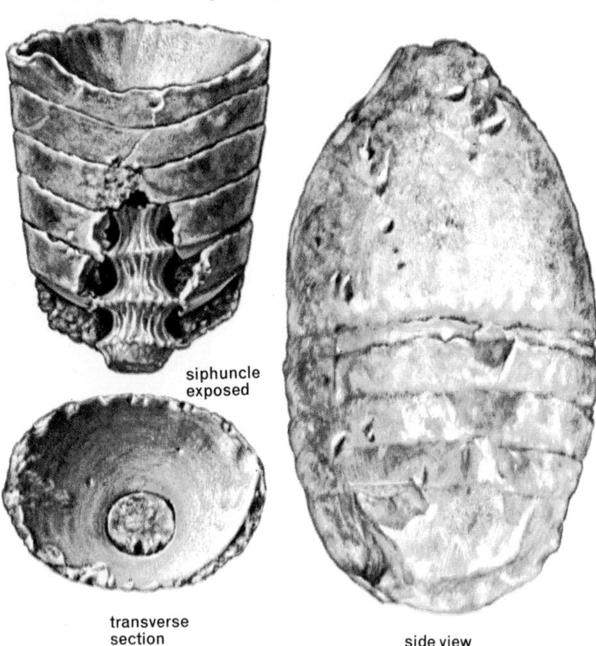

ENDOCERAS (Actinoceras), Ordovician. A group of long, more or less straight, conical shells, with a large calcified siphuncle (shown above), which has internal funnel-like structures, nautiloid sutures and prominent recurved septal necks. Max. length about 10 ft.

GOMPHOCERAS, Ordovician to Devonian, refers to a group of stout, bulbous nautiloids, with straight or slightly curved shells and a large body chamber. They had simple septa, T-shaped apertures and smooth or striated surfaces. Length about 3 in.

ing simple nautiloid sutures (p. 36). More tightly curved forms with sharply folded (ammonoid) sutures began to appear in the Silurian. After the Devonian the nautiloids persisted, but they decreased in numbers, while later ammonoids had wavy (strongly folded) suture lines.

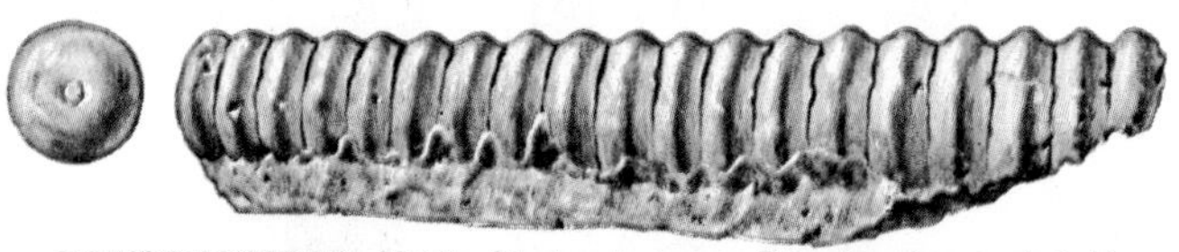

DAWSONOCERAS, Middle Silurian to Lower Devonian, has a straight conical shell, with a ringed surface and wrinkled growth lines. Small central siphuncle. Length about 5 in.

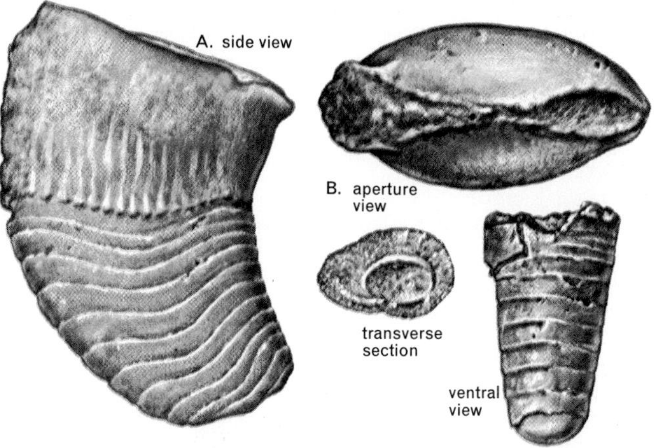

PHRAGMOCERAS, Silurian, has a strongly curved, laterally compressed shell. Aperture long but restricted and with a lipped margin (B) and figure 8-shaped outline. Siphuncle on concave side. Transverse striations. Length usually 4 to 5 in.

DOLORTHOCERAS, Devonian to Pennsylvanian, has a straight, conical, smooth shell, circular in cross section. The sutures are transverse and slightly sinuous. Has a central siphuncle. Feeble concentric or transverse ornament. Length about 4 in.

LATE PALEOZOIC
AND EARLY MESOZOIC CEPHALOPODS

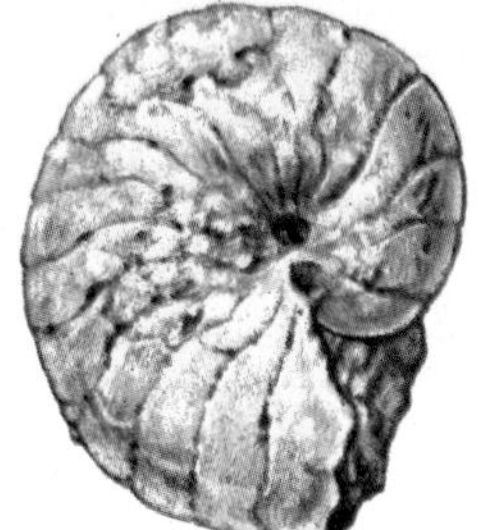
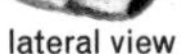
lateral view

ventral view

ventral view

AGONIATITES has a flattened, tightly coiled shell with faint growth lines. The ventral siphuncle goes through straight septal necks. Suture very simple, with ventral lobe. Middle Devonian. Diameter about 6 in.

BACTRITES has a straight and slender shell which is rounded in cross section. Sutures very simple, with small ventral lobe. The siphuncle is ventral. Possibly ancestral to ammonoids. Ordovician to Permian. Length about 1.5 in.

CYRTOCERAS is a short, curved, conical nautiloid. The shell is rounded in cross section and has a prominent ventral siphuncle. Ordovician to Devonian. Length 2 to 3 in.

GASTRIOCERAS has a shell which varies from globular to flat, with a prominent depression at center of the whorls, with a ribbed margin. Suture with simple primary folding. Pennsylvanian. Diam. 1.5 in.

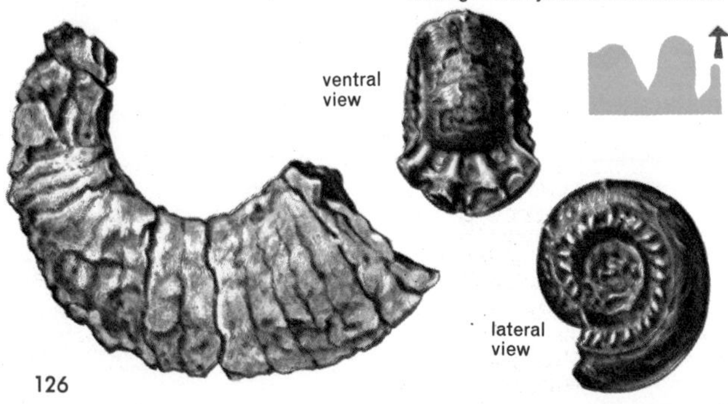

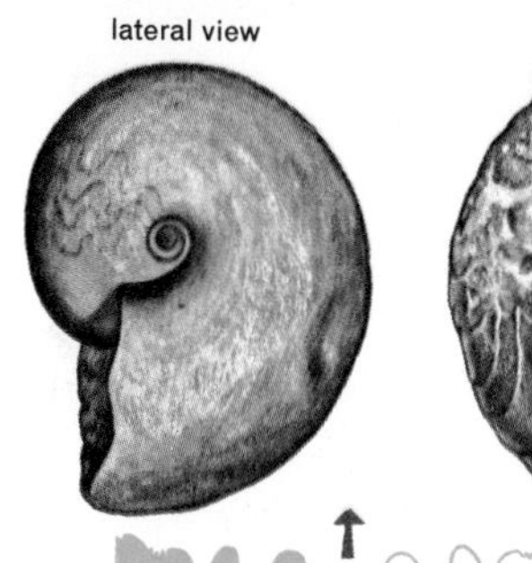

Suture patterns adapted from Moore, Lalicker and Fischer

Blue diagrams show wavy suture lines important in identifying ammonoids; red arrow points to aperture (p. 73).

MEEKOCERAS is a flattened cephalopod, generally smooth, with a flat exterior edge. The suture has some secondary folding in the lobes. Lower Triassic of Idaho, California and Asia. Diameter about 2 in.

MUENSTEROCERAS is a globular to flattened shell with a prominent depression at the center of the whorls. Suture has deep ventral lobe, with straight sides. Mississippian. Diameter about 1 in.

COLUMBITES is a flattened, tightly coiled shell with an arched outer edge. Ornament is feeble. It has a suture with minor secondary folding of the lobes. Lower Triassic. Diameter about 1.5 in.

GONIATITES is a cephalopod with a globular, smooth shell surrounding a small but prominent depression at the center of the whorls. Suture line very distinctive. Mississippian. Diameter about 1 in.

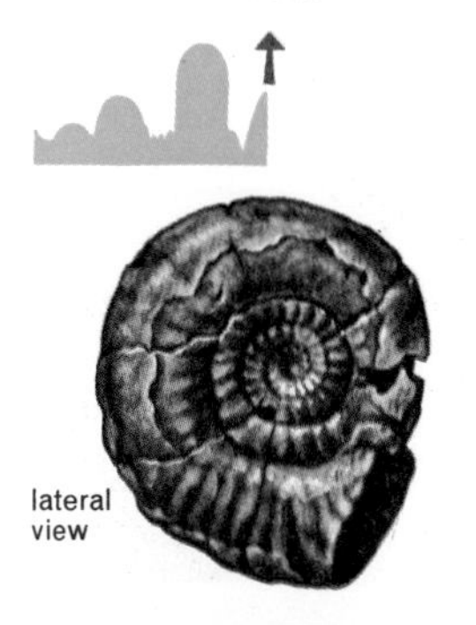

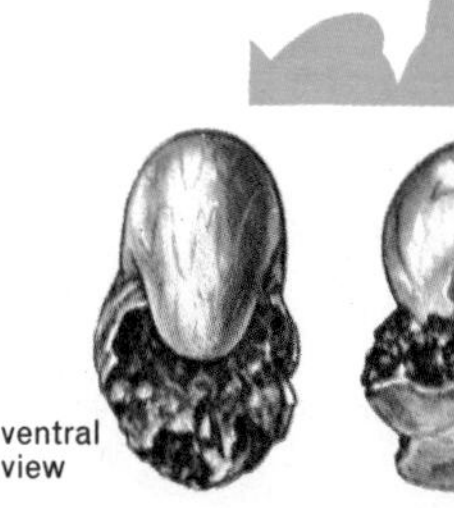

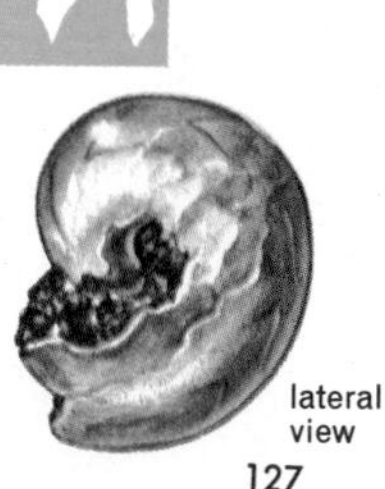

MESOZOIC

CERATITES, Middle Triassic, is a tightly coiled, robust shell, with the last whorl feebly embracing earlier ones. The coarse ribs do not extend to flat or broadly arched edge of the shell. Sutures are distinctive. Diameter about 2 in.

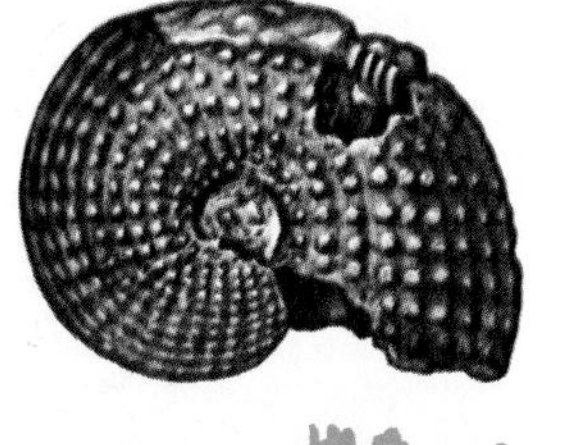

SAGENITES, Upper Triassic, has a globular but compressed, tightly coiled shell. It has spiral and radial ornament, the latter extending over the shell edges. Complex suture. It may have short spines. Diameter 2 to 3 in.

HILDOCERAS, Lower Jurassic, has a flattened shell that is somewhat square in cross section. It has three prominent ridges on the outer edges of the whorls and a wide central depression. Strong sickle-shaped ribbing on sides and complex sutures. Diameter 2 to 3 in.

HAMITES, Lower Cretaceous, is loosely coiled in one plane, with one short and two long shafts which are circular in cross section. The prominent ribs extend across the outer edge of the whorls. Complex suture lines. Length 2 to 3 in.

CEPHALOPODS

DACTYLIOCERAS, Lower Jurassic, a tightly coiled, flat ammonite with numerous whorls. Many ribs, the later ones branching. The ribs extend over the rounded outer edges of the whorls. Long body chamber and complex suture. Diameter 2 to 3 in.

PACHYTEUTHIS, Jurassic to Lower Cretaceous, is a belemnite with short, stout, blunt guard (slender in young forms). Fossil is suboval, eccentric in cross section and often large, with a groove down one side. Length 3 to 4 in.

STEPHANOCERAS, M. Jur., is thick and tightly coiled, with the last whorl feebly embracing the others; ribs prominent, continuous across the edge and branching at middle of whorl. Long body chamber. Aperture may have hood-like lips. Maximum diameter about 5 in., but usually smaller.

TURRILITES, Cretaceous, has a high-spired shell with whorls barely in contact. It looks like a gastropod, but is distinguished by presence of septa and a complex pattern of sutures. Conspicuous transverse ribs or tubercules. Length about 5 in.

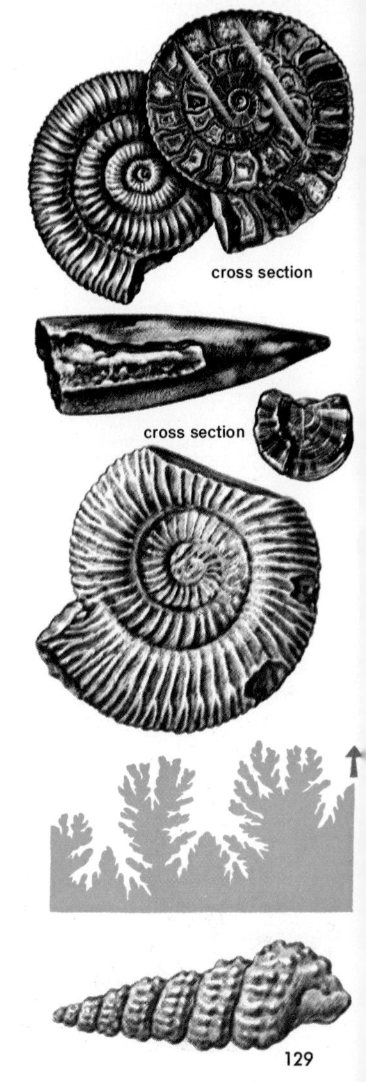

MESOZOIC CEPHALOPODS

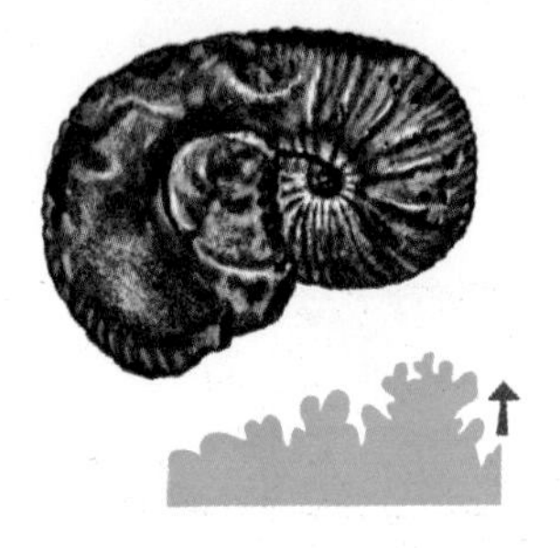

SCAPHITES, Cretaceous, is a flattened spiral coil with the early whorls in contact and the last one free, with short straight shaft and a hooked end. Ornamented with ribs which often branch; some bear tubercules. Length 1.5 to 2 in.

BACULITES, Upper Cretaceous, has a straight shell except for a small spiral initial stage. Surface smooth or with curved striae or low, rounded ribs. Suture symmetrical with intricate folding. Specimen's position in total shell shown above. Maximum length about 6 ft. but usually 3 to 6 in.

ACANTHOSCAPHITES, Upper Cretaceous, is a tightly coiled shell with the last whorl greatly expanded, and often extended into an oval shape. Prominent ribs and often nodes. Very complex suture pattern is characteristic. Diameter usually 2 to 4 in.

BELEMNITES, Miss.-Cret., are common Mesozoic cephalopods. They consist of long, bullet-like internal skeletons (guards), with a conical structure or depression at one end. One side of the guard may have a furrow and it may also have branched markings. Length 2 to 5 in.

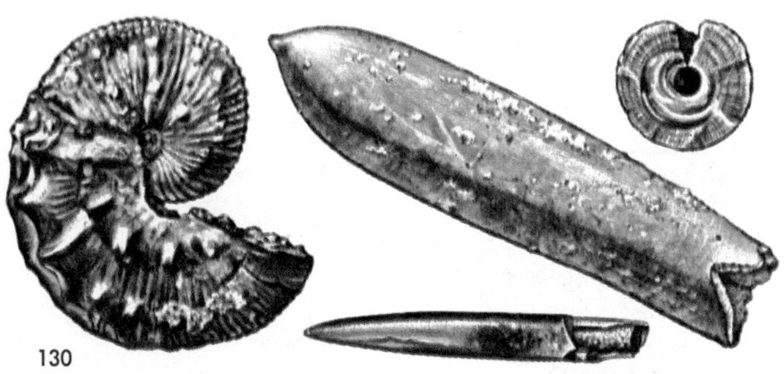

GRAPTOLITES are extinct marine, colonial organisms related to the protochordates, a group closely related to the vertebrates. The typical graptolite consists of one or more chitinous branches (stipes) bearing cup-like structures (thecae). See p. 74. They are important early Paleozoic index fossils.

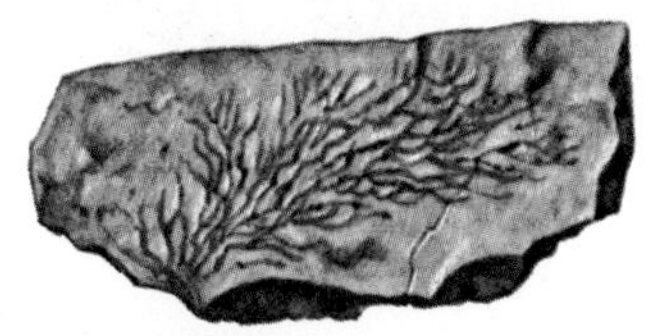

DENDROIDS, Upper Cambrian to Mississippian. Branching fan-like graptolites with numerous thin stipes. May have a root-like base. Maximum length about 4 in.

DIDYMOGRAPTUS, Lower to Middle Ordovician, have two stipes diverging at angles up to 180°, and each has one row of cylindrical thecae. Length of stipe 1 to 2.5 in.

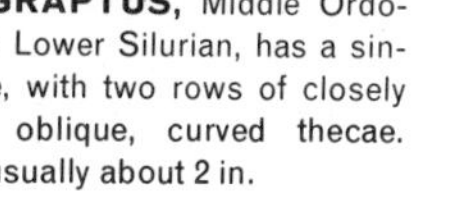

DIPLOGRAPTUS, Middle Ordovician to Lower Silurian, has a single stipe, with two rows of closely spaced, oblique, curved thecae. Length usually about 2 in.

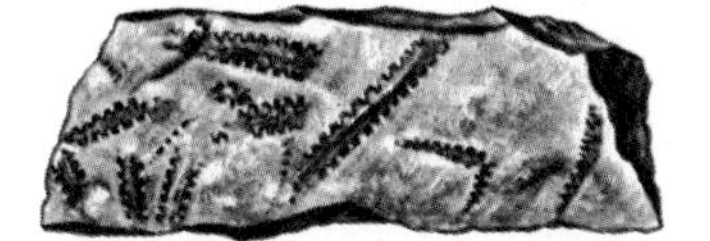

CLIMACOGRAPTUS, Lower Ordovician to Lower Silurian, has a single, straight stipe, with two rows of sharply curved thecae. Outer walls parallel to axis of stipe. Length 1 to 2 in.

MONOGRAPTUS, Silurian, has a single straight or curved stipe, with one row of thecae, which are very variable in form. Length usually 1 to 2 in.

NEMAGRAPTUS, Middle Ordovician, is an important and widespread guide fossil. The two S-shaped stipes give rise to numerous short branches. Length 1.5 in.

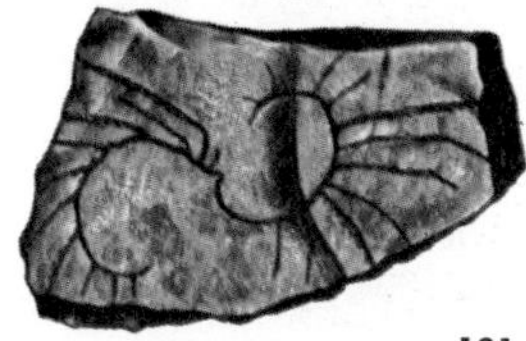

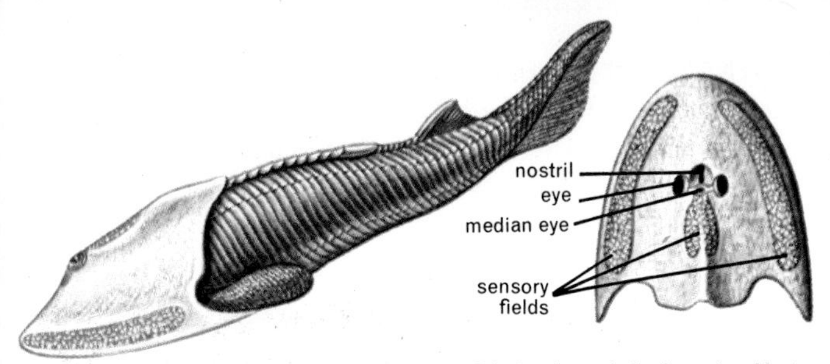

CEPHALASPIS is a typical member of a group of closely related ostracoderms (primitive, jawless fish), ranging from the Upper Silurian through the Devonian. Most cephalaspids had a flattened, bony head shield, and a scale-covered body. Length 6 in. to 1 ft.

VERTEBRATE FOSSILS

Vertebrates (fish, amphibia, reptiles, birds and mammals) and a few minor, primitive groups belong to the Phylum Chordata. All have a dorsal nerve cord with a supporting rod (notochord) later replaced by the vertebral column in vertebrates. All have gill slits if only at some stage of their life history.

Fishes, the oldest vertebrates, are divided into four classes: agnatha, the jawless fish; placoderms, the plate-skinned fish; chondrichthyes, the sharks and rays; and osteichthyes, the bony fishes.

The agnatha, the most primitive vertebrates, lack paired fins and true jaws. The oldest fossil vertebrates are Ordovician bony fragments from Wyoming and Colorado. Most fossil agnatha were ostracoderms, which were covered by an armor of bony plates or scales. They included both marine and fresh-water species. Both are widespread in Devonian strata. No agnatha are known in post-Devonian rocks. Their living representatives (the parasitic lamprey and hag fish) suggest the later forms may have been soft-bodied.

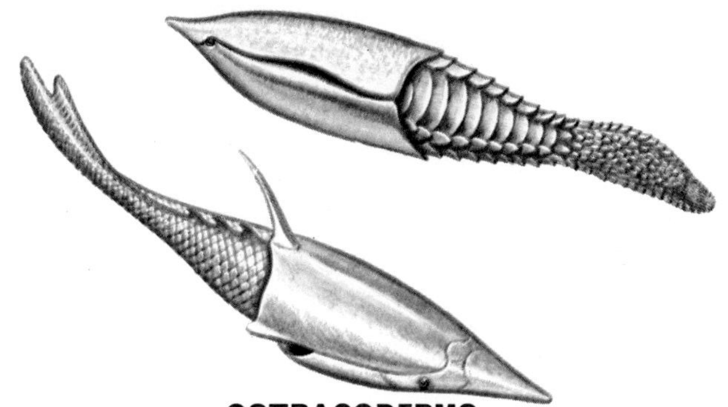

OSTRACODERMS

PTERASPIS, an Upper Silurian to Devonian ostracoderm, is a small streamlined fish. The head is enclosed in two large oval plates, with a rear spine and gill openings. Length about 6 in.

THELODUS, Middle Silurian to Lower Devonian. The entire body of this ostracoderm is covered with stud-like, interlocking denticles. Small lateral eyes; mouth below; flattened body. Length 3 to 8 in.

ANGLASPIS is a Devonian ostracoderm, has a large oval head shield, and widely spaced eyes. The trunk and tail are covered by distinctive scales. Length about 6 in.

DREPANASPIS, Lower Devonian, has a very large, flattened head shield of large and small fused plates. Vertebrate fossils are usually incomplete unlike whole specimens shown here. Length 1 ft.

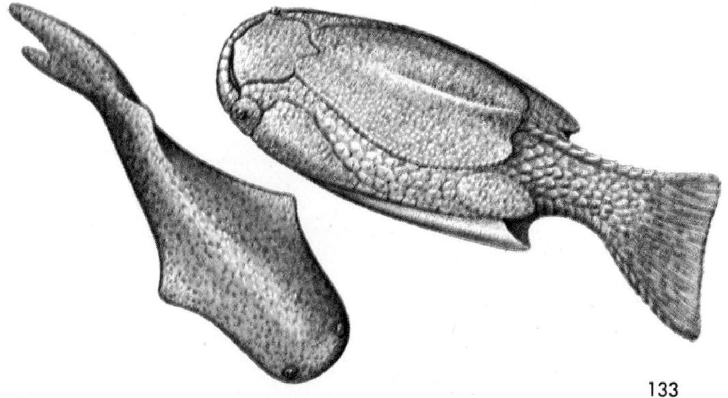

PLACODERMS (plate-skinned) are an extinct (Upper Silurian to Permian) class of fishes with primitive jaws and paired fins. A variety of forms includes acanthodians (spiny sharks), small, spiny-finned, streamlined, fresh-water fish, which were covered by thick scales; arthrodires (jointed-necks), with a heavily armored head and shoulder region, and wide gaping jaws; and antiarchs, small, armored, box-like bottom dwellers, with powerful, arm-like fins, abundant in the mid-Devonian.

PTERICHTHYS, Devonian. Front part of body covered with highly arched, fused plates; rear with scales. Closely spaced eyes. Strong "arms." Length 6 in.

BOTHRIOLEPIS, Devonian, an antiarch with front heavily armored by a short head shield and a long, box-like body shield. Long, jointed "arms." Length about 9 in.

COCCOSTEUS, Devonian, a jointed neck arthrodire with armored head and thorax. Rest of body naked. Exposed bony plates serve as teeth. Length 18 in.

CLIMATIUS, Upper Silurian to Devonian. A spiny acanthodian "shark" covered with rhomboid scales, 2 spines on back; 5 pairs of ventral fins. Length 3 in.

SHARKS AND RAYS (Chondrichthyes) have a skeleton of cartilage and open gill slits. Most are marine predators with well-developed teeth and are protected by bony scales. These and an occasional spine are usually the only fossils found. In contrast to placoderms, sharks have two pairs of paired fins, and more specialized jaws and teeth. The oldest sharks (Devonian) underwent great expansion in the Upper Paleozoic. Mesozoic and Cenozoic forms were widespread.

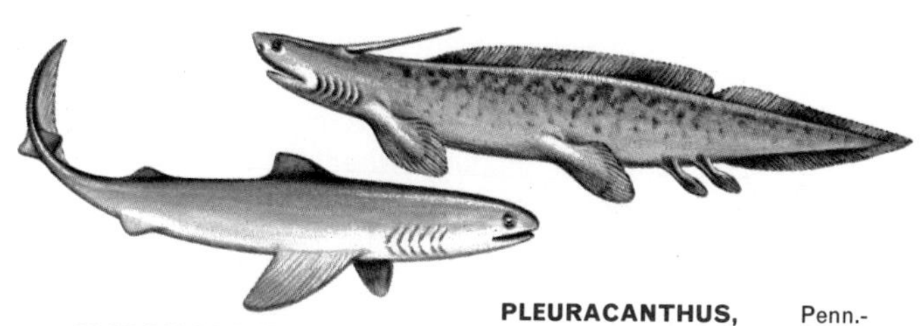

CLADOSELACHE, Upper Devonian. Well-developed, broad fins and streamlined, with naked body. Teeth numerous, pointed. Maximum length about 4 ft.

PLEURACANTHUS, Penn.-Perm., is a fresh-water shark, with greatly elongated dorsal fin and pointed tail. Paired, leaf-shaped ventral fins. Spine at back of head. Length usually about 2.5 ft.

SHARK TEETH are common fossils in some rocks of the Miocene. Largest are those of *Carcharodon,* a 40- to 50-ft. shark. Most fossil teeth are well-preserved.

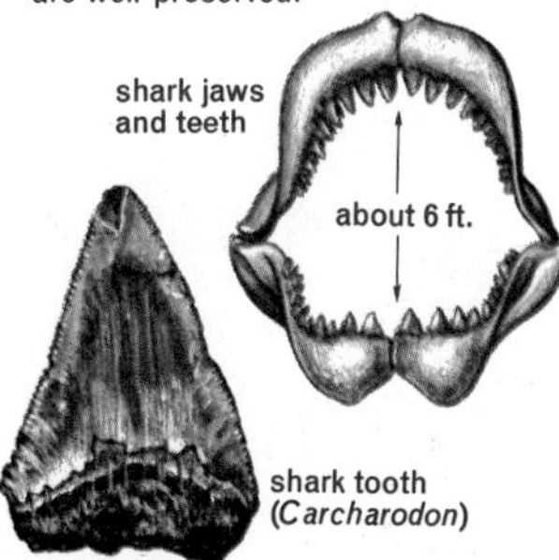

shark jaws and teeth

shark tooth (*Carcharodon*)

RAYS AND SKATES, Jurassic to Recent, are bottom dwellers, with flattened bodies, huge pectoral fins and heavy, shell-crushing teeth. Rare as fossils.

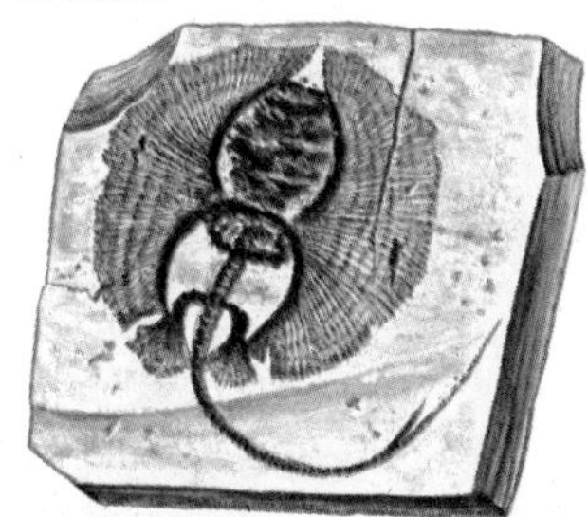

From a photo by Smithsonian Institution

BONY FISHES (Osteichthyes) are the most abundant, diverse and complex group of fishes. They outnumber all other fishes twenty to one and include more species than all other vertebrates combined. Bony fishes have a bony skeleton and slimy, scale-covered bodies. Some fossil and a few living bony fishes have lungs; the rest have an air bladder that controls buoyancy.

Most of the early bony fishes had streamlined bodies and well-developed fins. These features permitted active swimming with minimum disturbance of the water and contributed to the fishes' rapid expansion. Large eyes and mouths, aids in evasion and food gathering, also helped bony fishes to flourish in lakes, streams and seas. Specific adaptations to unique environments developed.

The oldest (Middle Devonian) bony fishes had thick enamel scales, which became lighter in later forms. In the ray-finned (actinopterygian) forms the fins are supported by many slender ray-like bones. Paleozoic ray-finned forms were a small fresh-water group, which later grew abundant in the seas. Only a few surviving fish, such as sturgeon, represent these primitive ray fins. Early forms were replaced in Mesozoic times by holeost ray fins, with more complex and efficient skeletons, jaws and scales. Surviving holeosts include the garpike and bowfish. In Cretaceous times most of the holeosts were replaced by the more advanced teleost ray fins, which include almost all living fish.

The other main group is the lobe fins (Choanichthyes). In this group, the fins are supported by a strong bony axis and the nostrils open into the mouth. Lobe fins include lung fish (Dipnoi—Devonian to Recent) and the fringe fins (crossopterygians). In turn, the fringe fins include coelacanths and the small, powerful, carnivorous fishes of the Devonian from which the amphibia arose.

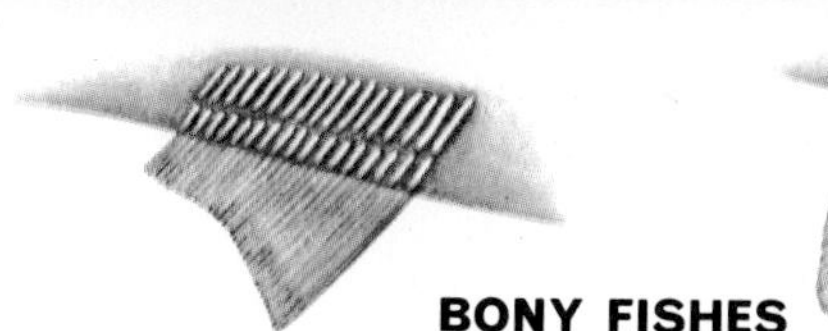

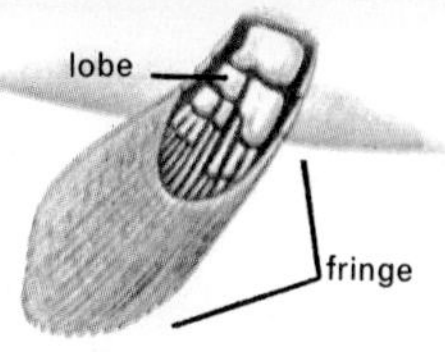

BONY FISHES

RAY FIN, showing typical structure of supporting bones, characteristic of most living fish.

LOBE FIN, showing typical structure of strong, supporting bones from which feet developed (p. 138).

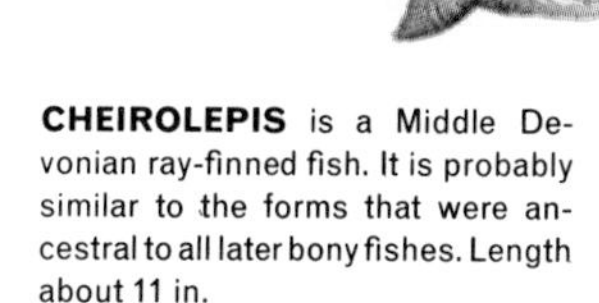

CHEIROLEPIS is a Middle Devonian ray-finned fish. It is probably similar to the forms that were ancestral to all later bony fishes. Length about 11 in.

OSTEOLEPIS, Middle Devonian, a primitive fringe fin (p. 138), with thick, rhombic scales; well-spaced median fins, short, lobed, paired fins and simple teeth. Length 9 in.

LEPIDOTUS, a Jurassic holostean ray fin, is a deep-bodied species with dorsal fin set far back and with two paired fins and anal fin below. It had strong crushing teeth and heavy enameled scales. Length 12 in.

HOLOPTYCHIUS is an Upper Devonian specialized fringe fin (p. 138), with the median fin set well back on the body and with long, lobed, paired fins below. Scales are rounded. Length about 2.5 ft.

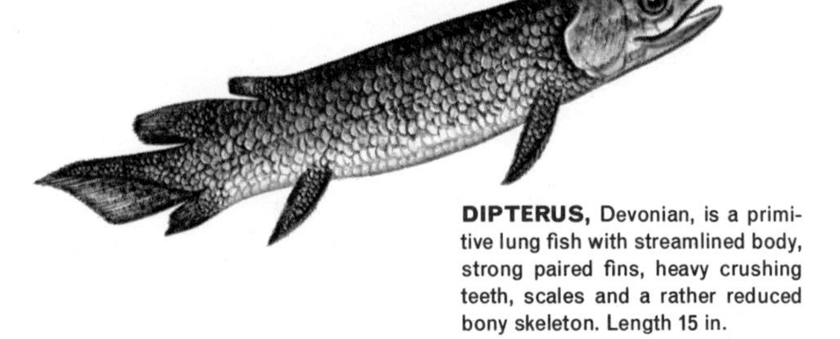

DIPTERUS, Devonian, is a primitive lung fish with streamlined body, strong paired fins, heavy crushing teeth, scales and a rather reduced bony skeleton. Length 15 in.

LOBE FINS, Choanichthyes, (lung fish and fringe fins) are air-breathing bony fish (Devonian to Recent) with internal nostrils and strongly supported fins. Three genera of lung fish (Dipnoi) survive. The fringe fins (crossopterygians), ancestors of amphibia, include living coelacanths and abundant Devonian forms.

EUSTHENOPTERON, from the Devonian, is a powerful carnivorous fringe fin or crossopterygian. It has an advanced structure and characteristics and is closely related to the ancestors of the amphibia. Length 2 ft.

COELACANTHUS, Miss.-Perm., is a typical deep-bodied, lobate-finned coelacanth, such as were common in the Mesozoic. Thought to be extinct, the first live coelacanth was caught off Madagascar in 1938. Length usually less than 1.5 ft.

ICHTHYOSTEGA, U. Dev., is a primitive amphibian, with many fish-like characters (bones of skull, vertebrae, tail), but others already amphibian (strong shoulder and hip girdles, limbs, ribs). Length 3 ft.

AMPHIBIA are the simplest tetrapods, the first class of vertebrates to invade the land. They are only partly adapted to land life, as nearly all lay eggs in water and larvae are gill-breathing, aquatic creatures that only later develop lungs and four limbs. Most living forms (frogs, toads, salamanders) live in damp environments.

The oldest amphibia (ichthyostegids) come from the Upper Devonian of Greenland and retain many characters reminiscent of their crossopterygian ancestors (p. 137). Most early amphibia were labyrinthodonts with infolded enamel in their teeth. These were sprawling carnivores, up to 15 ft. long. Mississippian to Triassic.

ERYOPS, Permian, is a large carnivorous labyrinthodont; heavy-bodied and squat with a large triangular skull; short powerful limbs. Well adapted to land life. Length 5 ft.

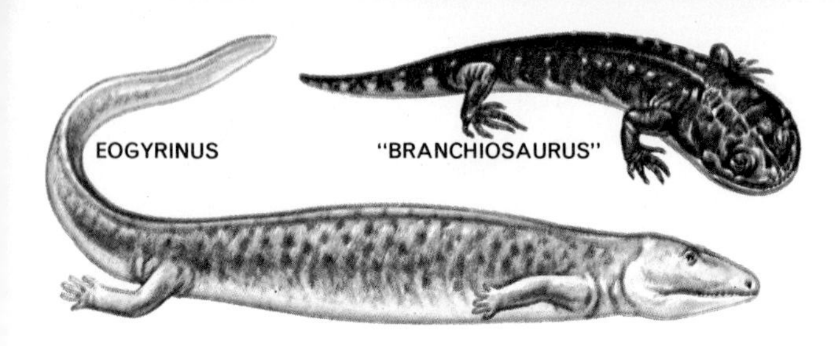

EOGYRINUS, Pennsylvanian. The deep powerful body and weak limbs of this very large salamander show it was largely aquatic in its habits. Length 15 ft.

"BRANCHIOSAURUS," Permian, are larval labyrinthodonts, with external gills and less bony skeleton than adults. Properly known by generic names. Length 2 to 3 in.

DIPLOCAULUS, Permian, is a wedge-headed salamander with weak legs. The spacing of the eyes and body shape suggest a life that was largely aquatic. Length 2 ft.

CACOPS, Permian, is a small land labyrinthodont with well-developed legs, armor plates on back, short tail and heavy skull. Length to 16 in.

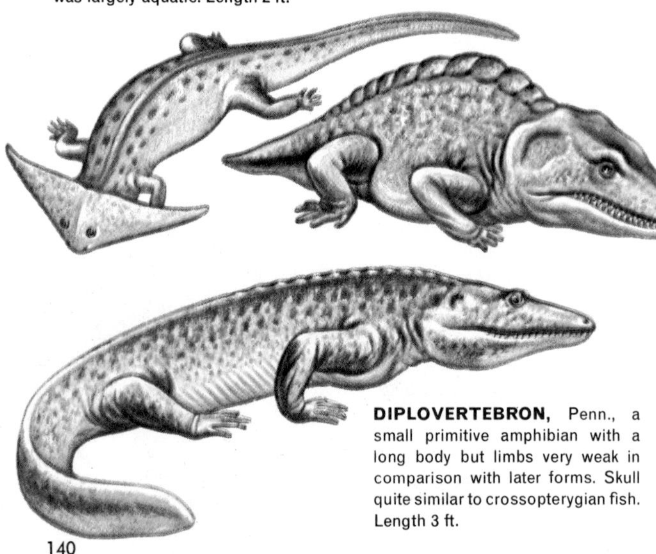

DIPLOVERTEBRON, Penn., a small primitive amphibian with a long body but limbs very weak in comparison with later forms. Skull quite similar to crossopterygian fish. Length 3 ft.

REPTILES (Pennsylvanian to Recent) are cold-blooded, egg-laying vertebrates including crocodiles, turtles, lizards and snakes. Fossil forms, more numerous and widespread, dominated Mesozoic life.

Reptiles are better adapted to land life than amphibians. Fertilization is internal; eggs are laid on land and include a food supply and a protective covering. Reptiles are protected by a skin with scales or plates. They breathe air through lungs.

Some of the earliest reptiles (seymouriamorphs) showed a mixture of amphibian and reptilian characters. Other early reptiles included pareiasaurs, pelycosaurs, mammal-like theraspids, and aquatic mesosaurs. The ruling Mesozoic reptiles included dinosaurs on the land, pterosaurs in the air and six groups in the seas.

Reptiles declined greatly at the close of the Mesozoic Era. Only four of fifteen major reptile groups still survive. All groups, extinct and surviving, have characteristic skull structures.

EDAPHOSAURUS a small-skulled vegetarian pelycosaur developed a vertebral "sail" like *Dimetrodon,* a larger-skulled carnivore. Compare them on p. 142. Permian.

SEYMOURIA, Permian, is a very primitive reptile, with many amphibian characteristics but with distinctive reptilian vertebrae and well-developed limbs. Length 2 ft.

DICYNODON, Permian, a therapsid. Length 7 ft.

CYNOGNATHUS, Triassic, a mammal-like reptile. Length 7 ft.

TRIASSIC

PERMIAN

MOSCHOPS, Permian, a "giant-headed" herbivore. Length 8 ft.

DIMETRODON, Permian, a carnivorous pelycosaur. Length about 8 ft.

EDAPHOSAURUS, Permian, a herbivorous pelycosaur. Maximum length about 11 ft.

OPHIACODON, Permian, a fish-eating reptile. Maximum length about 8 ft.

EVOLUTION OF THE MAMMAL-LIKE REPTILES
After E. H. Colbert

EARLY MESOZOIC REPTILES include a great variety of forms, most of which arose in the Permian from an ancestor not greatly different from *Seymouria*. Mesozoic forms include sea-going turtles, placodonts and ichthyosaurs as well as many terrestrial forms. Among the land reptiles were sail-backed pelycosaurs, mammal-like reptiles and small bipedal thecodonts, which were the ancestors of dinosaurs, birds, crocodiles, snakes and flying pterosaurs. The reptilian conquest of land, sea and air is one of the major events of earth history. No other group of animals except the mammals have shown such a range of adaptations to varied climates and environments.

MOSASAURS, Cretaceous sea serpents, probably evolved from lizard-like ancestors. Powerful, crocodile-like bodies, strong jaws and teeth, and well-developed paddles. Length 30 ft.

AMPHICHELYDIAN turtles, ancestors of present land and sea forms, arose in the Triassic. Some Cretaceous turtles reached 12 ft.

PLESIOSAURS, Jurassic to Cretaceous, were marine carnivores. They were active swimmers propelled by powerful paddles. Some were long-necked, with small heads and long tails; others were short-necked and long-headed. Length 15 to 40 ft.

ICHTHYOSAURS, or fish lizards, are carnivorous marine reptiles (Triassic to Cretaceous), with streamlined fish-like bodies. Well-preserved fossils show they gave birth to live young. Length up to 30 ft.

NOTHOSAURS are slim, fish-eating amphibious reptiles from the Triassic, related and perhaps ancestral to the plesiosaurs. Length 4 ft.

PLACODONTS are Triassic, mollusk-eating reptiles, with walrus-like bodies and highly specialized teeth. Some forms had bony armor on the back. Length about 11 ft.

After E. H. Colbert

MESOZOIC MARINE REPTILES dominated the oceans as dinosaurs did the land. Using paddles and lungs instead of fins and gills, these descendants of former land animals became highly adapted to sea life. Turtles and mosasaurs developed from different ancestral stock than other marine reptiles.

DINOSAURS, the terrible lizards, are the best known of all reptiles. They dominated life on the land during most of the Mesozoic Era, a period of about 140 million years. Dinosaurs arose from a group of Triassic thecodont reptiles and are represented by two main types. One group, the saurischians, had a reptilian hip structure. The other group, the ornithischians, were bird-hipped dinosaurs.

Saurischians developed into two distinct groups of dinosaurs. The most primitive were the theropods, the oldest of which were small (5 ft.), slender, bipedal creatures, with a long balancing tail. Later forms included the giant carnivores. The second group were the sauropods, most of which were large, four-footed, long-necked herbivores. The largest of these grew up to 87 ft. long. Teeth, sharp in carnivores and blunt in herbivores, usually extended the length of both jaws.

Ornithischian dinosaurs included four groups. Stegosaurs were 20-ft. herbivores with a high-arched, armored back from which heavy bony plates rose in a double row, and a spiked tail. Ornithopods were semi-aquatic, duck-billed, bipedal dinosaurs with webbed feet. The largest were about 25 ft. long. Some ornithopods had crested air-storage structures on the skull.

Ankylosaurs were armored, tank-like dinosaurs; with strongly curved ribs, whose broad backs were covered by overlapping bony plates, some modified into spikes. They were 20 ft. long.

Ceratopsians were horned dinosaurs ranging from 5 to 20 ft. in length, with thick head and neck armor. They were well-protected plant-eaters.

Dinosaurs were worldwide in distribution and inhabited a variety of different environments. The reason for their extinction at the close of Cretaceous times is not known.

SIMPLIFIED FAMILY TREE OF THE DINOSAURS

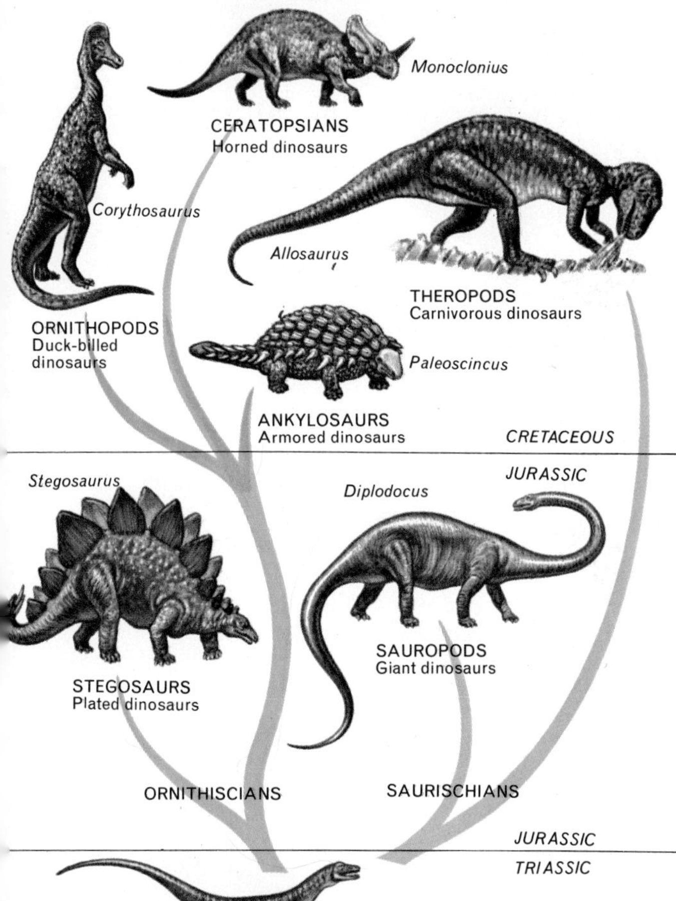

After E. H. Colbert

ARCHAEOPTERYX, Jurassic, a primitive crow-sized bird with many reptilian features. Known from the Solenhofen limestone quarries in Germany. 18 in.

BIRDS are winged, feathered, warm-blooded, egg-laying vertebrates. The oldest known bird, *Archaeopteryx,* shows many reptilian characteristics (such as teeth, clawed wings and a reptilian tail). Birds probably arose from thecodont (socket-toothed) reptiles, from which dinosaurs, crocodiles and pterosaurs also evolved. Their delicate skeletons and way of life make birds uncommon as fossils. Cretaceous forms include sea birds and divers. The Cenozoic includes a number of giant, flightless, carnivorous birds, some 10 ft. high.

HESPERORNIS, a Cretaceous toothed, loon-like, flightless sea bird, well adapted for swimming and diving. Note the small, almost useless wings. Maximum length about 6 ft.

PHORORHACOS, a heavy-billed, flightless Miocene land bird from South America. Height about 5 ft.

MAMMALS (Jurassic to Recent) are warm-blooded vertebrates that suckle their young. Most mammals have hair or fur, strongly differentiated teeth and highly developed senses. They are the dominant group of living animals the world over.

MEGATHERIUM, a Pleistocene ground sloth. Length 20 ft.

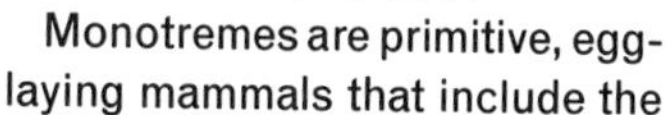

Monotremes are primitive, egg-laying mammals that include the duck-billed platypus and spiny anteater. No monotreme fossils are known before the Pleistocene, but they probably developed much earlier.

Marsupial mammals (kangaroos, opossums) have young which are immature at birth and are sheltered in the mother's pouch. Marsupials (Upper Cretaceous to Recent) were common in South America. Geographic isolation aided the development of this unique fauna there and also in Australia where they still persist.

Placental mammals (Cretaceous to Recent), the largest group, have an intermediate structure (the placenta) by which the embryo is nourished. Carnivores (dogs, cats, seals) are flesh-eating placental mammals. Fossil forms include creodonts, fissipeds and pinnipeds. Creodonts, archaic Tertiary carnivores, were mostly rather small, slender and long-tailed. Fissipeds, split-footed carnivores, replaced the creodonts. They probably arose from weasel-like ancestors. Modern fissipeds (dogs, bears, weasels and cats) evolved at various times during the Tertiary. Most have good fossil records. Pinnipeds, web-footed carnivores, include seals and walruses, which probably evolved from dog-like ancestors in Miocene times. Mammals range in size from 2-in. shrews to blue whales, over 100 ft. long, the largest known animals.

OXYAENA, Paleocene to Eocene, is a carnivorous creodont. Compare its structure with *Phenacodus,* below. Length 3 ft.

HOOFED MAMMALS (ungulates) are mainly rather large herbivores. They include living horses, cattle, elephants and hippos, as well as sea cows and a few fossil forms with claws rather than hoofs. The oldest ungulates from the Paleocene include small condylarths, with only partly modified teeth and either claws or very rudimentary hoofs, amblypods, up to 4 ft. high, with elephantine limbs, and the larger uintatheres. In Eocene times these forms were replaced by odd-toed, extinct titanotheres and chalicotheres as well as by primitive rhinos, horses and tapirs. Even-toed ungulates (deer, camels, pigs, cattle) appeared in the Eocene, and in late Tertiary times largely displaced the once abundant odd-toed ungulates. The horse and rhinoceros are the best known living odd-toed ungulates. The evolutionary history of many hoofed mammals is known in considerable detail (pp. 5 and 67); some are classics in paleontology.

UINTATHERIUM, Eocene, is the typical six-horned, herbivorous uinthathere, of interest because of its highly modified teeth. Length 12 ft.

PHENACODUS, Paleocene to Eocene. An advanced condylarth, but still retaining a long tail, five discrete toes and a carnivore-like skull. Length 6 ft.

MOROPUS, a Miocene, horse-sized chalicothere, with titanothere-like teeth. Its three functional toes are developed into strong claws. Height about 6 ft.

BRONTOTHERIUM, Oligocene. 8 ft. high. The largest of the titanotheres. Early titanotheres were slender, hornless creatures, only about 2 ft. high.

EDENTATES are a group of mammals with much reduced teeth. They originated in South America and later migrated into North America. Two main groups have developed. One, the armadillos, is protected by heavy armor over the shoulders and hips, joined by flexible plates. The oldest forms come from the Eocene, and related late Cenozoic forms included the giant glyptodonts with massive solid armor.

The other group of edentates includes the tree sloths, anteaters and the extinct ground sloths. Some of the great Pleistocene ground sloths were up to 20 ft. long. They fed on leaves and fruit browsed from trees.

NOTHROTHERIUM, Pleistocene. A small ground sloth, about 7 to 8 ft. long, was a contemporary of early man in the southwestern United States.

GLYPTODON, known from the Pleistocene, is a specialized edentate with fused bony armor and a heavy armored tail, some ending in a spiked club. Length 9 ft.

CRYPTOZOON "REEF", Cambrian. These and similarly shaped limestone features were probably made by simple algae.

FOSSIL PLANTS

Plant life has always been the basis for animal life but plant fossils are less numerous and plant life of the past is less well known. Note the marine and land plants in the dioramas from p. 34 to p. 69.

THALLOPHYTES

Thallus plants are simple, lacking roots, stems, leaves and conducting cells. Algae, thallophytes with chlorophyll, manufacture their own food. Of seven large groups, only a few are preserved as fossils. Some go back to Pre-Cambrian time. Thallus plants without chlorophyll include fungi, slime molds and bacteria. These have left even less of a fossil record.

CHARNIA (left), Upper Pre-Cambrian, is a disputed fossil known from England and Australia. It is regarded by some as an alga and by others as a sea pen—one of the coelenterates. Length 4 to 8 in.

DIATOMS, Cretaceous to Recent, are small unicellular algae, usually microscopic. They are free floating and have delicate siliceous skeletons. Diatoms occur in both fresh and salt water. These minute algae form deposits of diatomacéous earth as much as 3,000 feet thick. About 10,000 living species are known. Some of them seem identical with those of the Cretaceous.

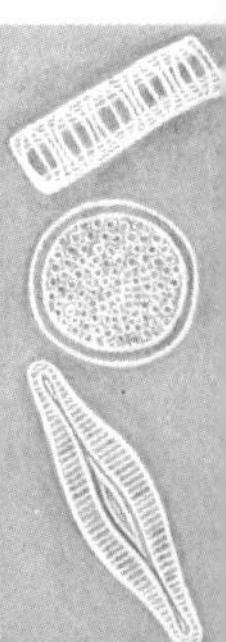

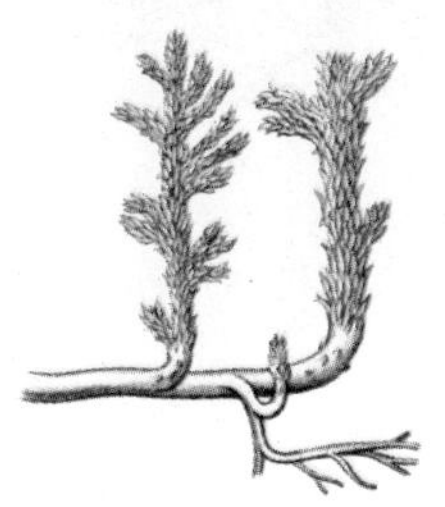

ASTEROXYLON, a Devonian psilopsid, has a simple branched stem, with leaf-like appendages. More complex than *Rhynia.* Length to 10 in.

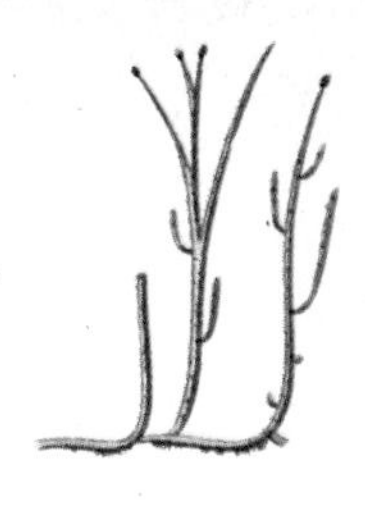

RHYNIA, a Devonian psilopsid, has a naked, branched stem bearing terminal spore cases. One of the simplest vascular plants. Length to 8 in.

EMBRYOPHYTES

In this large and important group of plants the fertilized egg develops into an embryo which is enclosed in some form of protective sac or covering. First in this group are the bryophytes, which include mosses and liverworts. These are the only embryophytes without specialized vascular tissues (see below). These simple land plants of moist places are rare as fossils. Pennsylvanian to Recent.

VASCULAR PLANTS

These plants contain specialized conducting tissues (xylem and phloem) usually with true roots, stems and leaves. Fossils quite common and widely distributed.

PSILOPSIDS, the simplest vascular plants, have small scale-like leaves or none at all. Roots are lacking. Common in the Devonian, they include the oldest known land plants from the Silurian of Australia. Only four species survive. Two fossils are shown above.

SPHENOPSIDS, or arthrophytes, include living horsetails and scouring rushes. These plants with ribbed, jointed stems and circlets of leaves bear spores in cones at tips or on stalks. Pennsylvanian forms (*Calamites*) grew 40 ft. high. Devonian to Recent.

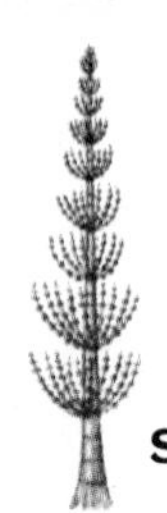

SPHENOPSIDS

CALAMITES, Mississippian to Permian. A scouring rush (sphenopsid) with ribbed, jointed trunk, and leaf whorls at joints. Ht. to 40 ft.

SPHENOPHYLLUM, Devonian to Triassic. Small with slender ribbed stems and circlets of fan-shaped leaves, 0.3 in. long pinnules.

LYCOPODS, Devonian to Recent, are vascular plants with simple leaves in spirals, never in circlets. Stem is not jointed. This group, which includes living club mosses, reached its zenith as large trees in the late Paleozoic. Some lycopods have two distinct types of spores. Fossils are common, especially associated with coal-bearing strata.

SIGILLARIA, Pennsylvanian. Stout trunk, with bladed leaves and vertical leaf scars. Maximum height about 100 ft.

LEPIDODENDRON, Pennsylvanian. Tall, branching, with slender leaves and diamond leaf scars. Maximum height about 100 ft.

STIGMARIA, Pennsylvanian; the roots of lycopod trees. Surface pitted with irregular spirals of rootlet scars. Longest are 40 ft. in length.

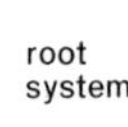

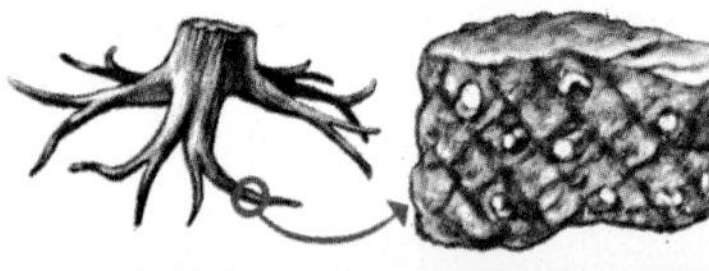

FERNS are an old group of vascular plants often with large, complex leaves that may bear spore cases on the undersides. They became common in the Pennsylvanian and are still common today. Many fern-like fossils are not true ferns, but seed ferns (below).

GYMNOSPERMS, the simplest seed plants, prospered because the male pollen grains resist drying, while in simpler embryophytes the male cell must be moist. Gymnosperms have no flowers and seeds are not fully enclosed. Living and extinct gymnosperms fall into five groups, illustrated on pp. 153-154. The seed ferns are an extinct group which developed seeds on their leaves, never in cones (Devonian-Jurassic). The Mesozoic cycadeoids and the closely related cycads had a ring of narrow palm-like fronds growing from a rounded trunk. The cycads differ from the cycadeoids in having the male and female cells in separate cone-like structures (Permian-Recent). The cordaites were tall trees with slender, strap-like leaves (Pennsylvanian-Triassic). Ginkgos (Triassic-Recent), common in the Mesozoic, include one living species. Conifers, mainly needle-leaved, with cones, are now most common. (Pennsylvanian-Recent.)

NEUROPTERIS, Miss. - Perm., frond of a seed fern with curved veinlets; oval leaflets alternate on either side of stem. Length of pinnules 0.25 to 0.5 in.

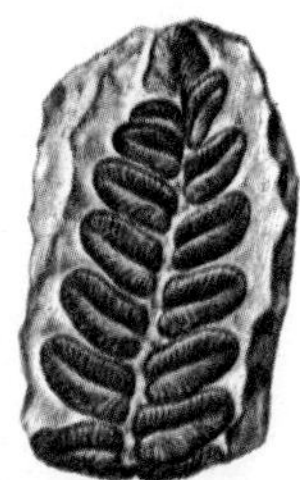

ALETHOPTERIS, Penn., another seed fern with long, blade-like leaflets, wider at the base. Mid-rib vein is very distinct. Length of pinnules 0.5 in.

GYMNOSPERMS

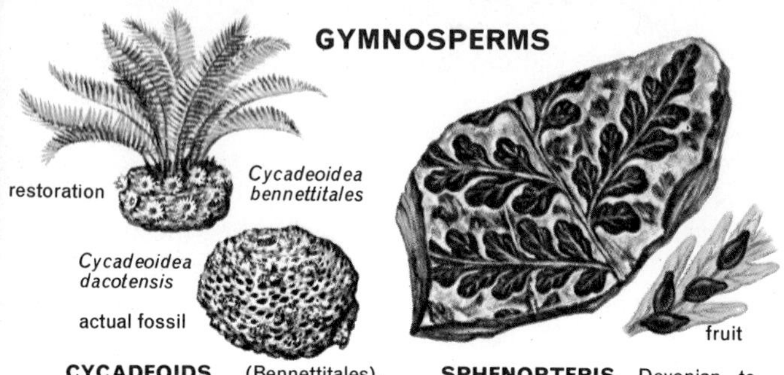

CYCADEOIDS (Bennettitales), Triassic to Cretaceous, dominant in Mesozoic, resemble living cycads but reproduction distinct. Height 2 to 12 ft.

SPHENOPTERIS, Devonian to Pennsylvanian. Frond of a seed-fern; small, symmetrical, lobed leaflets with radiating veins. Length of pinnules 0.4 in.

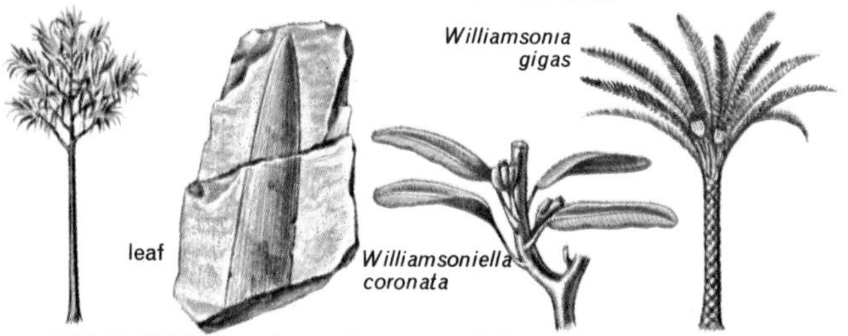

CORDAITES, Devonian to Triassic, grew worldwide in Late Paleozoic. Possibly ancestral to conifers. Maximum height about 100 ft.

LEBACHIA, a Pennsylvanian to Permian conifer, had a straight trunk and spirally arranged needle-like leaves. Branch shown about 10 in. long.

WILLIAMSONIA, Triassic to Cretaceous. Cycadeoid plants with bulbous stem covered with sunken leaf bases. "Flowers" on long stems. Height about 6 ft.

GINKGO, Triassic to Recent—a living fossil, with leaves on short spurs. Widespread in the Mesozoic. Leaves to 4 in. in length.

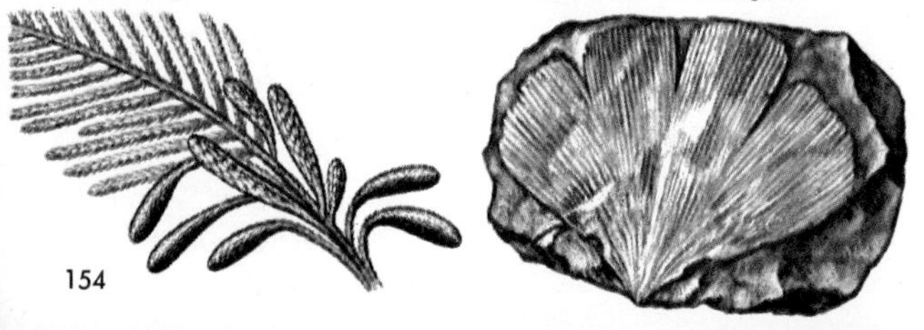

A Cenozoic landscape with angiosperms and some conifers; both very similar to modern trees.

ANGIOSPERMS are the flowering plants, which include about a quarter million living species. The flower is a unique organ. Pollen grains spread by insects or wind produce a tube by which the egg cell is fertilized. Seeds are enclosed and protected. Angiosperms developed in the Cretaceous and later became so important that the new grasslands favored the evolution of hoofed grazers such as horses, antelope and cattle. Angiosperms include the monocots (*Monocotyledons*)—grasses, lilies, sedges, palms, pineapples and orchids, and the dicots (*Dicotyledons*)—p. 156—with such families as the rose, mallow, mustard, buttercup, tomato, mint, carrot and daisy. Fossils of both groups are common in fresh-water clays, volcanic ash and other fine sediments.

MONOCOTS

GRASS, known chiefly as fossil seeds, became widespread in the Miocene and had a strong influence on mammalian evolution. 0.1 in. long seeds.

SANMIGUELIA, Triassic, a palm-like plant from Colorado. If it is a palm, it is the oldest known angiosperm. Maximum length of leaves about 1.3 ft.

DICOTS

MAGNOLIA, Cretaceous to Recent. A widespread and common but very primitive angiosperm. Grew in Alaska and Greenland in late Mesozoic and Cenozoic.

WILLOW (*Salix*), Cretaceous to Recent. Long (usually 3 to 6 in.), slender leaves; edges finely toothed. Common. Fossil pollen is especially important as a microfossil.

BIRCH (*Betula*), Cretaceous to Recent. Widespread trees. Leaves 2 to 4 in. long, oval, pointed, edges toothed. Common fossils indicating a cool, temperate climate.

FIG (*Ficus*), Cretaceous to Recent. This family, which includes many tropical fruit trees, is typical of warmer regions. Widespread in Cenozoic. Leaves 6 to 12 in. long.

SASSAFRAS, Cretaceous to Recent. Medium-sized trees or shrubs. Leaves 4 to 6 in. in length either simple and oval or with 1 to 3 bulbous lobes. Related to laurel.

MAPLE (*Acer*), Cretaceous to Recent. Medium to large trees. Widespread in temperate regions. Leaves to 12 in. in length, broad, lobed and toothed. Fruit winged.

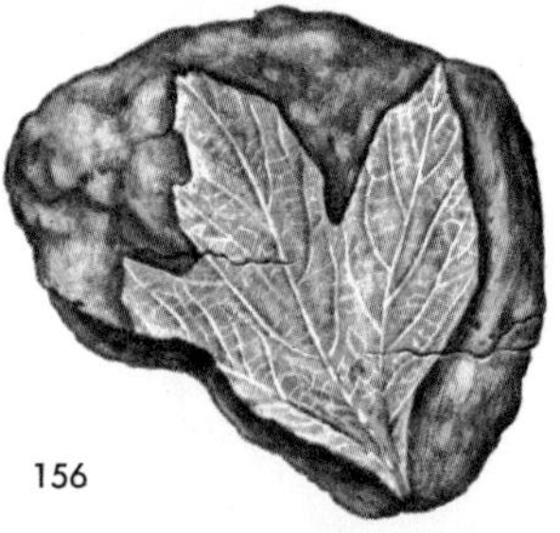

INDEX

Asterisks (*) denote items which are illustrated; **bold face** indicates major treatment.

MEASURING SCALE (IN MILLIMETERS AND CENTIMETERS)